PRAISE FOR

If the Creek Don't Rise

"In this astonishing memoir Rita Williams descends into the caverns of her past and returns with a tale no less dramatic than the rugged mountains where it takes place. A remarkable book by a remarkable writer."　—Janet Fitch, author of *White Oleander*

"Rita Williams's vivid prose introduced me to a fascinating, rarely seen corner of the American West."
　　　　　　　　　　—M. G. Lord, author of *Astro Turf*

"Williams is a gifted writer and she has crafted a fascinating story. *If the Creek Don't Rise* is highly readable, immensely entertaining and should be on everyone's summer reading list."
　　　　　　　　　　—*Tucson Citizen*

"Williams . . . is a gifted storyteller, and her tales of Daisy are unforgettable."　　　—*Publishers Weekly* (starred review)

"Enthralling . . . This work successfully brings readers into the world of the author's struggles and triumphs."
　　　　　　　　　　—*Library Journal*

"The descriptions of people and small-town life are dead-on."
　　　　　　　　　　—*Los Angeles* magazine

If the Creek Don't Rise

If the Creek Don't Rise

Rita Williams

A HARVEST BOOK
HARCOURT, INC.
Orlando Austin New York
San Diego Toronto London

www.HarcourtBooks.com

From *Tao Te Ching* by Lao Tsu, translated by Gia-fu Feng and Jane
English, copyright © 1997 by Jane English. Copyright © 1972 by
Gia-fu Feng and Jane English. Used by permission of Alfred A.
Knopf, a division of Random House, Inc. Photo of Portia Mansfield
on page 221 courtesy of the Tread of Pioneers Museum, Steamboat
Springs. The photo on page 51 was taken by Kurt Markus.

The Library of Congress has cataloged the hardcover edition as follows:
Williams, Rita (Rita Ann)
If the creek don't rise: my life out West with the last Black
widow of the Civil War/Rita Williams.—1st ed.
p.   cm.
1. Williams, Rita (Rita Ann)   2. African American
women—Biography.   3. African Americans—Biography.
4. Colorado—Biography.   I. Title.
E185.97.W73A3   2006
978.8'00496073092—dc22   2006000335
ISBN 978-0-15-101154-4
ISBN 978-0-15-603285-8 (pbk.)

Text set in Fournier
Designed by Cathy Riggs

Printed in the United States of America

First Harvest edition 2007
K J I H G F E D C B A

CONTENTS

The valley spirit never dies;
It is the woman, primal mother.
Her gateway is the root of heaven and earth.
It is like a veil barely seen.
Use it; it will never fail.

—LAO TSU

Don't go opening your mouth about things
that's none of your business. Hush up.
Hold your mud.

—DAISY

If the Creek Don't Rise

1 ~ Playback

*My sister Mary, me, Daisy, and my mother
in Strawberry Park*

~~~~ Out my kitchen window, the November wind
off the Pacific whipped up light frothy waves on Silver Lake.
The oddly beautiful smog-seasoned light burnished the last of
the lemons, making them look sweeter than they were. I was just
spritzing juice on a bowl of raspberries when my cat, Banana
Sanchez, cool from her dalliance in the garden, settled to warm
herself on the answering machine downstairs. The playback

button engaged. I heard a familiar gravid thro
breath and a sigh.

"Rita? This is Rose." A pause, as if waitin
up. I didn't. "All right, here it is. Daisy calle
called me and I'm calling you. Daisy say she fi:
to come." Still another pause, in which my ol
to be calculating how frank to be on the mach
what you aimin to do, but I ain't goin no plac

An abyss opened up right there over tl
Backed up against the Pacific, with two decade
my life in Steamboat Springs, Colorado, I s
numb fright at a summons from Daisy that I'
went to live with her at the age of four. The od
truly "fixin to die" right now were slim. She
that card for as long as I could remember. But I
I'd have to go back once more, as an adult. A
her nineties. Even she couldn't live forever. It

I called the airline and bought a nonrefund
I called the hospital and asked to speak to my auɪɪ.

"I'm sorry, ma'am," the nurse said. "Mrs. Anderson took
and checked herself out last week." There was no need to add,
"Against doctor's orders."

I had to laugh. Daisy had probably made those deathbed
phone calls to Rose and Mary from her home phone in Hayden.
Nothing terrified her more than being caged, and no pain was
more excruciating than the prospect of entrusting her body to a
doctor. She would have threatened to sue the nurses, the hospital, the janitors, the entire town of Steamboat Springs to get
herself out.

I dialed Daisy's number and waited, imagining her creeping
toward the phone. She picked up on the eighth ring. "Hell-*oo*?"

She still hollered into the receiver as if she had to make her voice carry all the way to California.

"Daisy, it's Rita."

"Reeter Ann Williams? Good land." Her voice sounded higher than it had six months ago, more pushed.

"Heard you needed some help. I'll be coming in Friday night late. Won't probably see you till Saturday morning."

"Well, I'll make up the bed and thaw out a rabbit."

"Yes, ma'am," I said. No need to start a fight right off by telling her that I intended to "waste good money" on lodging and a rental car. "I'll call you when I get to town."

"Well, I'm fixin to sell my books at the Christmas sale Saturday." I had to bite my tongue not to ask whether she meant to sell them from her deathbed, the priest standing by to deliver the final sacrament.

"Okay, I'll help you," I said, and I got off the phone, noticing that that old feeling had me in its spell, that sensation of sparring and dodging out of range.

~~~

I'd forgotten what it was like to fly into Steamboat Springs. The prop planes that looked so big on the tarmac at Denver International shrank against the fourteen-thousand-foot blade-sharp peaks of the Continental Divide. My breath grew shallow as the engines ground to the top of their range and the aircraft began to hop around like a waterdrop on a hot griddle. I always managed to fly in just ahead of a storm.

Only twelve other passengers could ride this little prop plane, mainly skiers accustomed to Rocky Mountain turbulence. Across the aisle, a kid in a ski hat was engrossed in an electronic toy—probably a snowboarder planning to spend Thanksgiving on the curl. A businesswoman was equally engrossed by the screen of her laptop. Only the working man with a white swath across his forehead where his tan ended and his cap began stared rigidly at the seat back in front of him, as if he too was holding his breath.

At last the pilot throttled back for the descent, and the plane took to bucking as if it had no intention of setting its wheels down anywhere near Steamboat Springs. I could not shake the image of a search party coming upon the wreckage of the aircraft and our bodies frozen among the pines of Berthoud Pass. Finally we broke cloud cover and the Yampa Valley lay below us.

My hometown was transformed. An infestation of shingled condos extended from the edge of the highway to the base of Mount Werner. The mountain face itself had been given over completely to the service of skiers, hundreds of wide gores scoring its flanks to create ski runs. I remembered when it was lowly Storm Mountain, debuting as a ski resort with a tattered little rope tow on a slope groomed by an orange snowcat. Daisy had been quite confident it would go belly-up in a season. But now I could see significant expansion on either side of Steamboat Springs. Wryly, I thought of an old sixties bumper sticker: Don't Californicate Colorado.

The plane lit on the icy runway, crow-hopped a little to the left, righted itself, reversed thrust. "Made it in before the blizzard," the pilot said over the PA when he'd brought us to a stop.

I walked down the stairs to the tarmac and rolled my little suitcase past what I could have sworn was the site of the A&W stand where I spent my saved allowance on my first root beer float. The arctic air clutched my throat. Hadn't felt that particular nip in a while. I hurried toward the tiny shed of a terminal, smaller than a single luggage carousel at LAX.

Inside, a wall was devoted almost entirely to snowboards. Where were the skis? But a surge of delight went through me as I saw the sensible face of a woman who had to be a Steamboat local behind the car rental counter. At seven thousand feet of constant sun and wind, people wrinkle early, and here was a woman like those I'd grown up with, more concerned with what she was doing than how she looked doing it. The wings on her eyeglasses hooked to the frame at the bottom—a style that had died out at least twenty years ago. Her hair had faded to the color of a weathered barn. She wore a tired white turtleneck under a plaid wool shirt, layers to protect her core. The label on her pocket protector read SUSAN. Something about the steadiness of her mien was familiar. Was she one of the Buchanan girls?

I was about to ask when a woman wearing a nuclear yellow ski suit and lizard-skin cowboy boots cut in front of me. "Did I leave my cell phone?" She set her plastic water bottle down on the counter and her acrylic nails clicked against the wood like bugs skittering on glass.

"Got any Jeeps left?" I asked, as if she hadn't spoken.

It was the boots that did it. You had to earn the right to wear cowboy boots like that. You had to scrounge around for years in mud and shit up to your ankles in cheap ones lined with cardboard that you tried to dry out overnight by the coal stove. Of course you had to wear them out damp the next morning. And you'd certainly wear them to show that Black Angus calf you'd hand-raised through Future Farmers of America, or 4-H. Daisy took me to F.M. Light & Sons when I was fifteen to get outfitted to help her at the county fair, and thirty years later I was still wearing the boots—olive-green suede with a wide heel and the subtlest swirl of scarlet stitching. Back then, their beauty was absolutely numinous to me, and that feeling never changed over the years as I traveled from New York to the high California desert. But in Colorado they were practical as well, in terrain that harbored Rocky Mountain spotted ticks, horse- and deer-flies, stinging nettles. I was the only black girl I knew who delighted in Garth Brooks's "Blame it all on my roots, I showed up in boots." I was still in sentimental love with the West, the romance of cowboy and horse, all those symbols of ennobled loneliness. And my secret sorrow would always be that I hadn't been able to make that life work for me. How was it possible that the membrane separating me from my redneck roots was still so thin that merely touching down on Colorado soil dissolved it?

Still, I restrained myself from telling neon-clad Ashley or Courtney or whatever-her-name-was to haul her ass to the end of the line and wait her turn. After all, I had long since thrown in my lot with the Botox crew in LA, and I was no stranger to a manicure. Susan had picked up on my tone, however, and we both stared at the interloper until she wilted. "Haven't seen your phone," Susan said curtly, then turned her full attention to me. "Welcome to Steamboat Springs, ma'am. How may I help you?"

It turned out she had only been in Steamboat for fifteen years, so I didn't know her. And they were out of Jeeps, nothing left but big sedans I'd never trust on a patch of black ice. She did have a Mustang in red. "That's my car," I said. I must have smelled right, because she gave me a deal.

The airport drive led out to Elk River Road, the way to the Sandelin Ranch where my friend Marie had lived. I'd ridden her family's big buckskin, Spook, on my first long trail ride to the upper range. I wondered whether the ranch house had become the site of a Holiday Inn. I felt a tremor shoot through a dam of long-suppressed feeling at the prospect of what I was embarking on. But I couldn't afford too much reverie with real weather bearing down. I took the turnoff for Highway 40.

The big urban-sized supermarket across the road surprised me. That meant the revenue base had gone up substantially, because I'd heard there was already a big Safeway out at Mount Werner. I decided to get provisions for the motel. When I got out of the car, I noticed how moist the air was. The snow, nearly upon us, would stick.

In the market I started at the sight of a little blond child, butterball plump, trailing a pout behind her wheezing mom, whose own knees had bowed under her bulk. They both looked inordinately pink in that industrial-strength fluorescent light. I couldn't remember any overweight people from my childhood here, certainly no kids so out of shape that a mere stroll down a grocery store aisle rendered them breathless. It felt odd to see the same hazelnut coffee, Fuji apples, and Earl Grey tea I would have seen on the shelves in LA, along with the leaf lard, flour, and pinto beans.

Back on the road, the November light that bathed everything in a pewter rinse right at dusk threw itself down on the valley as abruptly as it always had. Outside "Little Steamboat," as they called it now, a shamble of new buildings had been thrown up. The old Dream Island Motel and Trailer Court was still there, though, looking more run-down than ever. Daisy had cleaned there, and sometimes she'd dragged me along to help. I had to dust the night tables and dressers, replace the little soaps, and help her strip the beds. I remembered the strangled stuffiness of a room where the heat had been left on, the pungency that lingered on a pillowcase after someone with dirty hair had slept there, the fear that I would forever be attached to a rag, mopping up behind people who mysteriously thrived on some more luxurious plane than the one allotted me.

In spite of the cold, I opened the car windows to inhale the clean evening wind, let it scrub my face. Through the occasional break in cloud cover, I glimpsed the new moon, cool and steely between mounded cumulus moving in over the mountains. I had no doubt Daisy was looking out the window that very minute, sniffing at the moon. "See that ring," she'd say. "Snow flies tonight."

It felt odd to be riding this stretch without her at the wheel. I left the windows open, cranked the heat high to keep my feet warm, and turned on the radio. The trumpet of Miles Davis curled into the Mustang, something from *Bitches Brew*. That was a surprise. Back then, there had only been KRAI, offering country music and the stock market report—livestock, not Spyder funds.

Traffic lightened considerably as I headed west where the land was flatter—fewer sedans, more muddy pickups with rifle racks. Here and there I recognized a barn that had borne up through a half century of snow. The horses had frost on their winter coats, their breath condensing into visible clouds. How I had longed for my own horse back then.

As I came down the hill to Milner, a gust of wind slapped the Mustang and sent it into a skid. I nearly overcorrected until the old impulses kicked in and I turned into the spin, waiting for the axle to forgive me and straighten out. How rapidly and simply death could come. The metal taste of adrenaline on my tongue, I rounded a bend to a railroad crossing and had to smile as the Daisy in my head declared, "Train won't say a damn word. Run you down and keep on rolling." From below the trestle, I could hear the river, muffled under ice. It smelled delicious. The scent of horse must was on the air, and cattle too, the perfume of a pine fire burning in a nearby farmhouse. I had to admit I was quietly delighted to be home.

As I drove farther west, the snowy pastures stretched out smoothly as if they had been tucked in for the winter under a white satin blanket. Fewer farmhouses here; development postponed, at least for now. I recognized the striations of limestone and granite cut into a small hill, probably around the time my father and uncles were mining coal in Mount Harris in the 1930s. A number of black miners and cowboys had lived there. But few people, black or white, seemed to remember that African American Westerners had existed at all, which was one reason why Daisy was always pushing the two of us out "in the public," as she called it. I had no doubt that tomorrow night, if she could maneuver it, she would have me booked solid to be shown off at the Knights of Columbus, the Ladies' Recreation Club, and the Sew and Sew Club.

The snow, which had been falling almost inadvertently, as if from a down pillow leaking feathers, picked up as darkness began to exterminate the tepid gray light. I felt the headlights of the car being pulled toward a mesmerizing vortex. As I came over a gentle rise, I saw the MacGregor power plant in the distance, with its tall tower and flashing red warning light. On a field trip in sixth grade, I'd been so terrified by the din of its huge turbines that I'd cried. When the foreman of the plant put on a rubber glove and dramatically touched the hot main grid, I'd marveled anew at the senselessness of adults.

The WELCOME TO HAYDEN sign came into view, sending a ripple of apprehension through me. After fifty years in spiffy Steamboat Springs, Daisy had sold her property for a song, and run to ground along the railroad tracks in Hayden. From what I could see in the snow and dim light, the town hadn't changed much. The streets were still wide, as if the planners had expected an influx of traffic that had yet to materialize. Brick and stone homes still sported their large verandas with plastic flowers in loud colors hanging by the door—no frivolous remodeling here. The newest fad seemed to be foot-wide painted tin butter-flies attached jauntily to the sides of houses. I knew those touches would appeal to Daisy.

About a half mile into town stood the Hitchin' Post motel, its horseshoe-shaped parking lot full of pickup trucks, a dusting of sleet on their windshields. I'd sprung for the deluxe suite, counting myself lucky to get a room at all because hunting season had barely closed and the place was still booked almost solid.

The first thing that hit me when I opened the door was the smell of outdoor men, the kind who sat at the small Formica-covered table and cleaned their rifles and reloaded their bullets. The stove, the fridge, and the double bed all looked serviceable. I found a good seasoned cast-iron skillet in the top of the kitchen cabinet, and there was an old dull coffeepot to percolate on the stove—no fancy electric drip here. This place, plain as a milking stall, would be a cozy retreat after a day spent wrangling my aunt.

The wall heater clicked like an old lady's dentures before it wound itself up and puffed on. I could hear a man clear his throat in the next room. I hoped he wasn't a snorer. I unpacked the unworn wool sweaters I had knitted in California. In the yarn store I'd always bypassed lighter silks and cottons as if I'd known that this moment would arrive.

With dry, warm feet and a steaming mug of tea, I found nothing was quite as lovely as a full-blown blizzard in the Colorado Rockies. I lay in bed listening to the hiss of the snow spitting against the window, which needed some patching where the putty had cracked. I got up and tucked a towel along the sill. Muffled chains rattled as the occasional car went by, and a heavy road grader scraped the asphalt. Just before I fell asleep, I thought I heard a bull elk bugle. Dimly I wondered if that was possible, this close to town. And then the blizzard settled down for a proper roar.

~ ~ ~

I woke early, mysteriously overjoyed. I started coffee, still in my nightgown, then opened the door and got a fistful of fresh snow to eat. I knew Daisy would want me to have breakfast with her—fried eggs, fried venison, fried potatoes, fried bacon—not to mention two or three big biscuits dripping with melted butter and jam. Odd that I had never struggled with my weight until I moved to the coast and learned the virtues of watery white fish.

Out in the lot, most of the pickups were already gone, and the snow had drifted up past my car's hubcaps. The dawn was hunkering down gray as granite, refusing to be informed by the timid sun. I would need to get some yellow glasses to deal with the glare. The inevitable couldn't be postponed for long. I was certain that Daisy knew I was in town by now, driving a bright red Mustang. She would be scandalized when she saw I had no chains, and I hated that I secretly agreed with her.

Once I was dressed and in the car, I allowed myself a procrastinating detour down the main drag to the other end of town. The local ranchers' and farmers' co-op—a branch of the one we'd belonged to in Steamboat—was already open, taciturn mechanics in jumpsuits selling truck parts and farm supplies. How foreign they looked to me—although of course now I must look like an outsider to them.

When it couldn't be put off any longer, I searched out Daisy's street. Sure enough, there was the railroad track Rose had told me about. Even without directions I could have picked out Daisy's place. The house would have looked more prosperous had it not been for the jumble of old washing machines and car parts, coal ashes and lumber moldering in the yard. Uncle Ernest's old garden trolley leaned against a shed, although Ernest himself was long gone. When his knees had gotten bad, he'd fastened a tractor seat to the wheels of a grocery cart so he could ride up and down the rows of his prize-winning vegetable patch. He was religious about weeding. Like his sister, he'd babied his glorious garden but made do with slave-quarter lodgings. Ernest had been a tall man, and pushing himself along on his cart, stopping to hoe here and there, he'd looked like a huge praying mantis, all dark skinny legs and arms, capped by a huge velvet sombrero. Now the trolley looked the very picture of dejection, gone to rust with a muffin of snow on the seat.

The windows were a giveaway too, steamed up like someone was holding a sweat lodge inside. First thing in the morning, Daisy always fired the stove so hot the chimney glowed.

I parked the car and checked in the rearview mirror to be sure my lipstick wasn't too bright. Then I got out, went through the gate, and knocked on the door.

"Who is it?" Daisy called out, low, although I was absolutely certain she had been peering through the curtain from the time my car turned onto the block.

The clatter of keys. Daisy always padlocked everything, though everyone else in the Yampa Valley still left their homes unlocked. Finally she got the door open, and I was looking down at a faded old woman in an incongruously new steel wheelchair. It seemed to me that a wave of panic had come roiling up, though to which of us it belonged I wasn't sure. There was certainly no mistaking the odor of urine that the Pine Sol had failed to suppress, however. And there was something else—a sweet smell that was all too familiar, and that never failed to make my gorge rise. Daisy had mixed up a batch of powdered milk that morning. I swallowed, breathing through my mouth. Some part of me said, *Turn around. You cannot save her. Go back to the car. There's still time.*

"Mary?" Daisy smiled widely. "Mary Loretta, is that you?" Even now, thirty years after she had made her escape, Mary, the favorite, was still the first of us who came to Daisy's mind. I was poleaxed already, caught between the desire to smart off and anticipation of the stricken look that would surely provoke. "Niggers always got to be fighting," she would say, recoiling.

"It's Rita, Daisy," I said. "Not Mary."

"Well, come in the house then. Don't stand there and waste up all the damn heat."

My aunt's shoulders had slumped with time, but the blue ring around her irises was as pale as ever, and she looked alert and ready for business. At ninety-one, her face was still almost completely without wrinkles. But her hair had grown thin and grizzled, and her lipstick had been applied with hands more accustomed to managing a posthole digger.

"Whose big fancy car is that?" Daisy asked, jutting her chin toward the Mustang. She looked me up and down, taking in my floor-length red coat from Morocco, my matching boots, gloves, and hat.

"I came in it," I replied, not quite answering her question. She looked up at me and smiled, as if to say, *Oh, I see. You're going to be cute.*

Hot as the house was, Daisy had safety-pinned herself up in a big wool sweater. On her pigeon-toed feet she wore scuffed white running shoes with no socks. The skin of her naked ankles was cracked like it had been freezer burned.

"I brought you a gift," I said, handing her a paper sack. The sight of her gnarled hands saddened me. Arthritis had bent her fingers sideways and swollen the knuckles until they looked like burls. How did she manage to hold things, clean herself, get around? As she looked inside the bag, she grinned, and as always, the part of her that delighted in surprises broke my heart. Daisy's face when she genuinely laughed belonged to a completely different woman.

Inside, the room was long and narrow, with a small table along the window and the kitchen at the far end. An older woman I hadn't noticed at first sat at the table, her gray hair held back with plastic barrettes.

"That gal's Annalee," Daisy said offhandedly.

"How do," said Annalee, smiling girlishly. She had mascaraed blue eyes and a surprising, wounded beauty that seemed to manifest itself against the odds, like a cantaloupe vine growing in a compost heap. I knew the exact rack at the drugstore where she'd stood figuring whether she could afford both the pink and purple barrettes.

"Howdy." I nodded, my western twang snapping back like it had never gone at all.

I looked around the room, taking in Daisy's old oak clawfoot table and Gama's rocker. The roasting heat emanated from the same potbellied stove that had been in the basement of our old house. At the end of the room, next to the deep freeze, someone had installed an electric stove. I doubted Daisy used it. She didn't believe in wasting electricity on modern contraptions. Next to the stove she had positioned an old TV tray with instant coffee, a bowl of sugar, and a box of powdered milk, an intensely unwelcome reminder of my childhood.

Daisy leaned forward in the wheelchair, and I realized that her pants weren't pulled up all the way. Was she planning to go to the sale like this? I stifled a groan.

She took the shrink-wrapped box out of the bag I'd given her and rolled herself over to the table. "Girl, hand me the scissors," she said to Annalee. The elaborate pink box with its cameo inset looked wildly out of place on the faded red checkerboard oilcloth. Daisy put on her reading glasses, which were so grimy I couldn't imagine how she could see anything through them. She peered at the packaging. "White Shoulders. Ain't this the prettiest thing?"

"I tried to find Jungle Gardenia," I told her. "I guess they don't make it anymore. But this is pretty close." She propped the box in place, and, fending off my offer of help, stabbed at it with her scissors until she had the amber bottle out of its wrapping.

Annalee leaned forward as Daisy sprayed the back of her hand. "Good land, that sure is beautiful." The three of us inhaled. I didn't like to admit how much I loved the old-fashioned, flashy scent too, although the perfume's name always gave me the impulse to hide the bottle from company. I imagined that for Daisy, it only increased the cachet.

Even White Shoulders wasn't a match for the scent of powdered milk and my revolting associations with it, though. Suddenly I felt so light-headed I feared I would fall. "Daisy, I need to go to the car for a moment."

"Don't you go no place a-tall," she said. "You messed up. Should'a been here for breakfast. I done told you we had to leave outta here by seven thirty. Sale starts at nine thirty and I got to get that first booth by the door." Her dentures slipped down the surface of her gums as she spoke, and she clenched her jaw in an effort to keep them in place. I guessed she hadn't bought new ones since I'd seen her last.

"Now, listen here. You gon' take me and her." She flung her hand toward Annalee as if she were a crate of onions. "Mrs.

Moore gon' take William after she picks him up from the bus station." Uncle Billy was the one who'd told me that Daisy had lost her mind, selling the property in Strawberry Park and moving to a shack by the tracks in Hayden.

"Go make up a elk sandvidge for yourself so you don't go hongry. It'd be a crying shame for you to go spendin every nickel you got on expensive food up to the fairgrounds." I definitely didn't want that door to open, so I obediently made my way to the kitchen.

Daisy remained transfixed by the road outside the gate, leaning up as high as she could while Annalee occupied herself rearranging the objects on the table, smoothing out the paper sack and box the perfume had come in. The same idea of order prevailed in this house as in the house where I'd grown up. Daisy believed in keeping everything out in case she needed it—loose screws, needle and thread, margarine tubs, receipts, bobby pins, Anacin, sugar bowl, reading glasses, spare change, melting butter, used tinfoil that had been washed so she could use it again, Three-in-One oil for the whetstone, and, most curiously, a .22 bullet.

In spite of myself, I had to admit that the garlicky elk smelled good. I tucked some into a bag along with a hunk of golden cornbread slathered with strawberry jam. I still couldn't for the life of me figure out how to make cornbread like Daisy's. Mine always hardened like Sheetrock.

"It's fifteen minutes before eight. I'm fixin to be late. I pays Mrs. Moore onct a week. You'd think she could get here on time."

"I got here early," said Annalee.

Daisy continued staring out the window as if she hadn't spoken.

"Daisy said you was a dancer out to dance camp," Annalee said in her chewy Southern accent. She was referring to Perry Mansfield, the summer camp where Daisy had cleaned when I was little. "I done dancin," she continued, to no one in particular. "I done tap dancin, ballet dancin, adagio dancin, and acrobatics. Acrobatics ain't dancin but similar." I was beginning to see what Billy liked about her.

Daisy didn't seem interested at all in Annalee's talents. On the wall behind her, a large poster of her first husband, Robert Ball Anderson, serenely surveyed the proceedings, with gentle eyes and a long beard that Daisy said he'd kept since he escaped off the plantation back in slavery days so his brother would recognize him if they found each other again. Mr. Anderson, as we called him, was the reason for Daisy's reputation as a historical figure and authoress. Ex-slave, Union man, buffalo soldier, he'd also been the most prosperous black pioneer in Nebraska history. His legacy included a two-thousand-acre ranch in the western part of the state that spared his kin the detour to northern cities that many blacks took as they migrated from the South in the early twentieth century. In all his long life, though, he'd never learned to write, so Daisy had written down Mr. Anderson's life story for him, in 1927. In the 1960s she'd republished the book, and decades later, she was still at it. At every public occasion she sold and resold it, with new addenda, each one crazier than the last. Hence her anxiety to get to the sale today.

"Annalee's sweet on William," Daisy said slyly. "She think he got money. Don't know he don't aim to give up a penny to these ol' womens runnin after him." She glanced at me. "Sit down. Rest your coat. No tellin how long we got to wait."

"I ain't sweet on nobody," said Annalee, pushing out her chin.

"'Cept dogs," Daisy said pointedly. "Just look at that trailer full of varmints." She leveled her gaze on Annalee, who blushed a hot cherry red and squirmed in her chair. "So help me God, ain't got but six hundred seventy-nine dollars and twenty-seven cents comin from the welfare oncet a month and spent two hundred nineteen sixty-three on a damn dog got hit by a truck. Already got four cats and one come up big. Gal swear she know the cat was made so she can't have babies. Cat had five baby cats. Now she got to feed nine cats, two dogs, and a rabbit. Won't fry up the rabbit neither. Ain't got the sense God give a goose."

Annalee laughed as if Daisy had made a good joke. All the screws sorted now, she carefully placed them in the margarine tub and began on the bobby pins.

I perched on the edge of the couch. Daisy was trying to maneuver her dentures over to the side of her mouth so that she could run her tongue over her gums, a habit that made me clench my teeth. I was also struggling with an almost irresistible impulse to fix her lipstick. A trickle of sweat ran down the middle of my back. I scanned the room, and my eyes lit on the box of Daisy's books by the door.

"Would you like me to take those out to the car?" I asked.

"That's Mrs. Moore's job," Daisy replied. "Caint depend on nobody. Didn't none a my family 'mount to nothin. I give 'em everything. Didn't nobody give me nothin. I thank the Lord jus' the same." How could that same old line still make me so furious, yet so desperate to figure out what would please her?

She tilted her head, and I had the sense that she was about to grill me. "You got to urinate?"

It had been decades since anyone had asked me that question, and the last person who'd done so was Daisy. "If I need to, I'll manage," I said, as calmly as I could, avoiding Annalee's eyes. She made a point of plucking at a loose string on the tablecloth.

Oblivious, Daisy continued. "If you need to urinate or move your bowels, best do it now." She pointed toward a closed door down the hall. "Ain't no telling when we'll be back. Course you won't be comin back here, will you? Well, you better get a rag and put some Clorox on it to take with you. Ain't no telling what you gon' catch up to the Hitchin Post."

"Daisy—" I strained to keep myself from rising to the bait like a bigmouth bass. "I just needed to have a place where I could sleep and know I wasn't bothering you."

"Bull," she snorted, not even bothering to throw me a glance.

A log broke apart in the potbellied stove. I looked around, desperate now for a serviceable excuse to get outside. Mr. Anderson met my gaze, dignified and benign. I wondered what he had been like at seventy-nine, and what it had felt like to marry this pistol of a woman nearly six decades younger, and what it had been like for Daisy to marry him.

A powerful low rumble set the house to trembling and the windowpanes to rattling, as if we were in the midst of a seismic event. When I looked out the back window, a bright yellow Union Pacific locomotive was bearing down the track just across the road. "Daisy," I said, "that train's awful close." But she didn't hear me. A second diesel came on, stroking hard behind the first, black smoke staining the pristine sky.

I bolted up.

"Where you going?" Daisy mouthed, but I kept moving.

Outside, I counted the boxcars rolling past, but lost count after thirty. They trailed as far as I could see. From that million-dollar acreage in Strawberry Park, how had she wound up here?

It had begun snowing again. Big, tender flakes floated down, catching on my eyelashes and tickling my cheeks. I stuck out my tongue. I had forgotten how much I loved winter. At least I could pick up a shovel and clear off the walk. What a surprise it was to find my heart thudding with the effort. The years in California had thinned my blood, acclimated me to sea level. How embarrassing. We'd always laughed at wheezing city slickers who keeled over stepping from the street up to the sidewalk. I took off my coat and tossed it on the porch, and ate a mittful of snow to cool myself down.

Daisy tapped her cane on the window, commanding me back inside. I pretended not to notice. She opened the door.

"Reeter Ann," she yelled, "get in here. I got to tell you what to do."

I heard the muted clicking of chains in deep snow. A pickup pulled up, with Uncle Billy riding shotgun. He was still as dark as chicory, but his favorite cap, which announced that he was PROTECTED BY SMITH & WESSON, had been reduced nearly to shreds. I imagined it was camouflage for his receding hairline. We hallooed awkwardly and then out popped Mrs. Moore, a pretty, white-haired woman with a spring in her step.

Once inside, she headed straight for the sink, collecting dishes along the way.

"We ain't got time for that," Daisy said, hoisting herself to her feet. Her pants barely covered her rear. I could feel my jaw tighten. She was going to embarrass me no matter what I did. Still I said, "Shall we pull up your pants before we get started?"

"Get out of my way, girl. We got things to do and my pants don't make no nevermind. I'll get on this coat." She picked up the coat, which lay on a TV table by the door. Mrs. Moore appeared to be sizing me up, to see if I could handle Daisy any better than she could.

"But Daisy, the coat's too short," I said. Her arms barely made it into the sleeves, the fabric was so constricted across the shoulder. But she was determined not to ask for help, and she struggled until she got it buttoned askew, so that it hung at a jag.

"All right, Annalee, go git in Mary's car," she said.

Nobody moved. Condensation streamed down the windows.

Daisy twisted around as best she could, trying to whirl on Annalee. "Didn't you hear me talking to you?"

"I thought you said her name was Rita," ventured Annalee, now putting on her coat as well.

"Girl," Daisy flared, "Rita's who I told you to ride with."

"All right, Rita, start up the car," chimed in Billy, not to be outdone. "Don't you hear her talking to you?"

By the time I had turned the car around, Daisy had made it through the door, but the two steps down took her so long, and the snow was falling so hard, that she looked like a big piece of gingerbread sprinkled with coconut when she got in. Finally we loaded the books, the wheelchair, and the walker, and everybody got into the correct vehicles.

"All right, girl, now don't you drive me too fast," she said. I crept along at fifteen miles an hour, encountering no traffic at all. "You don't know how to drive in these conditions and somebody will fly up on you and kill you in a minute."

When we got to the barn that housed the sale, not a single car was in the parking lot. I tried the front door, but found it locked. The rodeo rink across the road, where barrel races were held in the summer, looked forlorn. Beyond the rink, pastureland rolled all the way out to the power plant. I got back in the car next to Daisy, who fidgeted and worried, craning this way and that, trying to conjure someone to let us in. Then Mrs. Moore pulled up alongside us, and she and Uncle Billy got out. "Say," Daisy hollered, "go see if you can get in at the side."

Mrs. Moore looked at Daisy for a good two-count, evidently trying to decide whether it was worth the effort to persuade her just to wait. But Daisy jumped her. "Go on. See them tracks?" The newly fallen snow had almost covered them completely. "I'll bet there's somebody in there right now, getting my booth. Got the front door locked so they can take their time."

Uncle Billy was out of Mrs. Moore's car now and ready to unload. "Daisy, we got plenty of time." He watched as Mrs. Moore followed the footprints around to the side of the building, and a minute later, she called to us that the door was open.

Daisy looked at me in exasperation. "I'm surrounded by mentals. All of 'em ought to be in a home. Now, let me out of this damn car."

As if on cue, one of the men I had seen earlier at the co-op opened the front door, and we went in, lugging the books. The ceiling inside had been hung with red, white, and blue bunting, and sound echoed off the hard walls as we dragged tables together. Daisy claimed the spot she wanted, settling in like an old hen. Annalee licked the pencil she was going to use to keep track of who bought books, but Daisy hung on to the cigar box where she kept her money.

As the day brightened, people whose faces I vaguely remembered began to materialize. I wandered the room. I hadn't seen crocheted toilet paper covers in a long time, or home-canned catsup, corn relish, or spiced watermelon pickle. There were fine handmade quilts too, and some beautiful simple yarn from a black sheep that left lanolin on my fingers. It would knit up into a sweater that would keep me warm and dry no matter how frosty the weather.

When I got back to the booth, Daisy had corralled a woman carrying a pan covered with aluminum foil. "You going to give a poor old lady one of them cinnamon buns?" Daisy whined musically, like a country western singer.

The woman smiled good-naturedly. She wore a beautiful blue Pendleton plaid shirt and jeans and high boots. "Good gracious, Daisy," she said, taking me in. "Is this that itty-bitty niece you used to bring in the market?"

"Sure is," Daisy said. "Got a master's dee-gree and been on the television," said Daisy. "Mrs. Hinson, give William one of them buns too, won't you?"

Mrs. Hinson pulled back the foil. Huge coils of yeasty brown dough, dripping with cinnamon, brown sugar, and butter and sprinkled with walnuts, smelled like absolute heaven on the cold air. "Rita, go on. Get you one too."

I shook my head, torn between my suddenly watering mouth and my disapproval of the way Daisy had ambushed this old lady. Daisy, undoubtedly sensing the reason for my hesitation, looked at me as if I were still her least favorite dimwit.

"Well," said Mrs. Hinson, starting to move off, "I got to be settin up my heating plate. Don't, these rolls'll be cold when folks starts coming." She winked at me. "You can come by and get you one later if you want."

"I'll do that," I said.

2 ~ Strawberry Park

My mother, Mae

~~~~~ **By the time I reached** the age of reason, Daisy was the only member of my clan left for me to ask about my family. Meandering down memory lane had no appeal for her, and she didn't like it when I pried into the past, any more than she appreciated other people asking, as they frequently did, why we wound up in the Colorado Rockies, at an elevation of eight thousand feet, the only black family for nearly two hundred miles. Why hadn't we stayed with our own kind? "Ignorant

sons of bitches," Daisy fumed. "Why in the Sam Hill don't they ask the white people why they don't all pile up in a little slum like a nest of field mice in a box full of old rags? Nobody questions why they come to live in the North, South, East, or West."

When I pressed, her answers were as variable as her moods, and often she refused to answer at all. Sometimes she told me that Gama, her mother, had had thirteen children, only five of whom survived to adulthood. Uncle Billy said there were only eight to start with, but when I pressed him about what had happened to the missing three, he shut down too. This much seemed clear: They grew up in a cabin in Tennessee, with little access to basic hygiene, medicine, education, information, or safety. Certainly Daisy, the eldest, felt Gama's burden the most, especially because all too often Gampa could only find work as a laborer on the Mississippi and Tennessee river steamboats. And Daisy had to shoulder a large share of the burden of raising seven brothers and sisters with almost no resources. This experience might have led her to any number of bitter conclusions, but the one she came to, as she repeatedly told me, was "Don't never take nobody else's child."

The reality of "freedom" in such a setting seemed tenuous. Forty years after the end of the Civil War, there was little for the white people to do except torment the people around them who had not so long ago been their slaves. Nightfall often found a gang of white boys creeping around the cabin, eavesdropping and generally playing the bogeyman to the terrified, ignorant, neglected children. Once in the middle of the night, Daisy said, the owner of the plantation where my grandfather sharecropped rode up to the cabin door and hollered, "Alice, you got to come." My grandmother was needed to wet-nurse the baby of a white woman who either couldn't or didn't want to nurse her own child. Having just had a baby of her own—my uncle

Reuben—Gama had milk. "Wasn't nothing she could do," Daisy said. "She had to go on out and climb up behind him on that horse's back, and there was no telling when we'd see her again." The baby she left behind lived only fifteen days.

Somehow Daisy managed to learn to read and write well enough to become a schoolteacher. At twenty-one, she caught the eye of visiting Civil War veteran Robert Ball Anderson. A former slave, Mr. Anderson had served both in the Union army and as a buffalo soldier in the 125th Colored Regiment of the Grand Old Army, where he might very well have been instrumental in helping herd the Colorado Utes down to their reservation in the late 1800s. Later he homesteaded in western Nebraska. By the time he had met Daisy, he had gone broke twice before finally amassing more than two thousand acres of viable ranch land and becoming the largest black landowner in the state.

At seventy-nine, Mr. Anderson felt that he could finally afford a companion, and after a monthlong courtship Daisy married him and moved back with him to his ranch, forty miles from the Wyoming border. After a few years, she returned for her parents and four surviving siblings. On another trip down South, she recruited more "field hands," one of them a ginger-colored man named Charles Lee Williams.

Initially Daisy took her brothers and sister Grace to live on the ranch with her, but arranged for my mother, Mae Ella, who was the youngest and her favorite, to live with a Jewish family in the town of Hemingford, about twenty miles away, where she could attend high school. There my mother took up the violin, which she learned to play well enough to win a scholarship to the University of Nebraska—no small feat for a black woman in the 1920s. Mae seemed to be on the verge of realizing the kind of life Daisy would have wanted for herself. But unbeknownst to Mrs. Anderson, as Daisy was now called, the handsome, ginger-colored hired hand named Lee had "commenced to courtin" Mae when he drove into town to bring her out to the ranch on Fridays. Perhaps Daisy didn't notice what Mae was up to because her attention was focused on expanding her husband's empire, a job for which she was ill equipped.

Despite having been born into slavery, Robert Ball Anderson had been his master's favorite, and had learned how to run a large operation from accompanying him about his Kentucky hemp plantation. And having gone broke twice, Mr. Anderson knew down to the jackrabbit what it cost to run his two-thousand-acre ranch.

Extremely frugal, he didn't even purchase lumber for a house. Instead, he lived on rabbits in a sod house for years because he knew that ten years of locusts, grasshoppers, and drought could easily wipe out the profits of ten years of mild conditions. But he was no match for his headstrong bride, who was in her visionary period. Impatient with his old team of oxen, Daisy invested in everything from new tractors to Percheron draft horses and blooded cattle. Somehow, however, she couldn't quite find the funds to pay Lee and her other workers—not unlike the landowner her father had sharecropped for in Tennessee. One of the more insidious lessons Daisy seemed to have learned in the South was that no one would investigate injustices toward black people. Maybe, never having had much money of her own, she simply didn't know how to manage a business. Apparently a banker in town was extending mortgage after mortgage to her as she made up for lost time.

Parcel by parcel, in much the same manner as it had originally been assembled, the ranch began to dissolve. Mr. Anderson's grain and livestock were sold off incrementally, without his knowledge. He stopped participating in town functions, no longer serving as color bearer for the Grand Army of the Republic in Hemingford's annual Fourth of July parade. The whispering about town was that he had become the virtual prisoner of his headstrong wife and her brothers. In 1930 Mr. Anderson died in a car accident that some people said looked suspicious. Shortly after that, the well-insured home that Daisy would ever after refer to as a "mansion" caught fire. And reportedly, when the firefighters arrived to put it out, my Uncle Ernest warned them off.

"Don't go near it," he was quoted as saying. "There's smallpox in there."

There were other rumors about Daisy's "friendship" with a local banker, but whether they were true or not, that didn't prevent his foreclosing on the Anderson ranch. "We had a thousand acres in potatoes and couldn't sell a one," Daisy said about those last days in 1929. Meanwhile, bypassing Daisy, Lee had asked Mae's parents for her hand in marriage, and they had consented. Daisy would consider her sister's decision to forfeit her violin, her scholarship, and her future for marriage to a hired hand a betrayal on both their parts, and she would never be able to forgive my father. Finally, she took off for South Dakota, leaving her parents to be supported by Mae and Lee. When Lee found a position on a cattle ranch in Wyoming, the four of them left Nebraska together.

Daisy was always reluctant to discuss her time in South Dakota, but my father and mother seemed to do well in Wyoming, with the exception of one species of wildlife that they encountered there. Rattlesnakes were plentiful, and one hot August day, Daddy raked up three of them along with the hay he was baling. He completed his shift, but while he was unharnessing the team that evening, he decided that between Arkansas and Wyoming, he'd had enough snakes to last a lifetime. When he got home, he told my mother, "Mae, I got to go where there's no rattlers." The rancher was sorry to lose him but advised him to head for the high Rockies, where there was lots of big game and, at seven thousand feet, nothing rattling. And so, around 1937, Mae and Lee and their two daughters came to Strawberry Park, Colorado, following the path along which the Utes had migrated in the fourteenth century.

~~~

Strawberry Park wasn't really a park. It was more of a clearing in the midst of the Rocky Mountains, with Copper Ridge to the west, Steamboat Springs to the south, Hot Springs to the north, and nearly five feet of dark topsoil underfoot. Millions of years ago, a glacier had gored this valley out of the mountains surrounding it. Despite wintertime temperatures that plummeted to fifty degrees below zero and an average annual snowfall of sixty inches, the brief summers were balmy. And because the growing season was so short, the soil never became depleted. Protein-rich timothy hay thrived in it. So did strawberries, as a resident by the name of Lester Remington found out in the early 1900s. One day out in the mountains, he discovered a berry so huge it would not fit in a water glass. He began cultivating strawberries, and discovered that at this latitude their forty-five-day growing period meant they did not ripen until late August or early September, well after strawberries at lower elevations had stopped setting fruit, and that they could survive a five-day trip to Denver in an unrefrigerated railway car. Had the First World War and competition from California growers not intervened, his business might still have been thriving by the time my family came to Strawberry Park.

For my people, who had come off the sparse plains of Nebraska, and before that out of the sharecropping South, this land was incomprehensibly generous. Like the Utes who had summered there before them, they found the forests full of game. The last gray wolf in Colorado had been killed by the time I was born, but there were bear and beavers, grouse and ducks, magpies, snowbirds, and owls.

At first, my parents found work picking potatoes. But my mother was a capable cook, a skill she used to supplement their income on various ranches during haying season. My father soon put his experience with horses and carpentry to work, and also got hired on at a coal mine twenty miles away, in Mount Harris. In little more than five years, they were able to purchase twenty acres north of town.

From their mountain foothold, they had plenty of opportunities to put to use the tracking and fishing skills they'd learned from living the subsistence life in the South. They were nearly first in line for runoff from the snows, and in the evening, the deer waited in the stand of aspens behind the house for them to go to bed, so that they could come down to the creek to drink. When darkness fell, the bugle of elk and the call of owls echoed down the valley. The creek itself provided German brown and rainbow trout, and the surrounding lakes were stocked. Along with those marvelous strawberries, there were currants and thimbleberries, gooseberries and serviceberries, chokecherries and raspberries—all for the martens to binge on until their mouths turned blue. My father said anybody with a pinch of industry could thrive in the valley, so great was its bounty. At first Mama made jams and jellies to sell from all that amazing fruit. But after she made dinner for some locals who raved about her way with a shortcake, the idea for a restaurant was born.

Right away, the Rushing Water Inn was a success. The restaurant served lovely fresh meals outside, next to the creek. My father attributed its success to my mother's culinary talent.

Her way with venison tenderloin, wild spinach, and wild onions surprised locals and tourists alike. But nothing compared with Mama's fried chicken, fresh garden vegetables, and strawberry shortcake.

Then Mama started thinking about the property across the road. What if they built some cabins for people to use as a base for hunting and fishing excursions? Daddy could build the cabins when he wasn't working at the mine, and they could bring Daisy out to help run the restaurant. Guests who stayed at the cabins would eat at the inn. Daisy, who apparently wasn't having an easy time of it in South Dakota, was only too delighted to come join her sister's going concern. Before long, their parents and their brothers Billy and Ernest had joined them as well.

I arrived in May, when the does drop their fawns. It was the season to be born in Colorado, offering warmth, abundant food, and just enough time to bulk up before the first frost, which sometimes launched winter as early as August. I remember the gentle sun on my face as I lay on an elk hide in a big field of snow, entertained by the roiling Colorado sky against the mountains, the racing thunderheads piling up on top of each other, lit from within, like mounds of freshly sheared wool. Lady, my black border collie, was not a bad babysitter. More than once she fetched me back from the edge of the creek where I had crawled for water so cold it gave me a headache. I must have crawled into the pasture next door on at least one occasion too, because I recall looking up at the whiskers and knobby knees of the mare our neighbors kept there, and her mild alarm at my presence. I don't remember feeling that I was in danger except once, when I had toddled out to the henhouse to listen to the odd high mumbling of the chickens, and the door slammed, shutting me in. The rooster—a huge white leghorn—attacked me from behind, his spurs goring my neck and cheek. Over and over, he rose up in the air, wings spread, red comb inflamed, diving for my eyes. The spanking I got for wandering off by myself doubled my sense of outrage at the existence of such merciless malice in the world.

I thrived on the secret life of plants and animals, which spawned in me a fundamental curiosity about the nature of things that would later become the gateway to my learning about objectivity. I created delectable mud pies for my dandelion dolls, whose long green hair I curled by splitting the stems apart and dipping them in the icy water of the creek. If only I had known there was such a thing as a scientist! Later, when catastrophes began to plague my family and me, the sight of cranes taking their rest at the bend in the creek before they continued their fall journey south brought me comfort even in the worst grief. Such elegance they had, like tall Africans visiting from latitudes where warmth and order prevailed.

~~~

In pictures as in memory, Mama was a pretty woman, her Cherokee and African heritage burnishing her skin the color of cinnamon tea. Her eyes had an upturned cast, and there was a softness to her, a sweetness hard to square with Daisy's assertion that she could put a bullet through the heart of a bull elk half a mile away. Behind the huge stove in the kitchen of the inn, her movements were rapid but calm, her face still, her gaze utterly excluding me. A sheen of perspiration glistened on her forehead as she eased the floured chicken into a skillet of bubbling grease. The big oven door clanked as she pulled out a pan of fragrant shortcake the color of buttered gold. Later she'd slice it open and pile on strawberries like shocking valentines aloft a scoop of stiffly whipped cream. Sometimes it was raspberries, and those I liked even better.

I was told my father was so handsome in his youth that women would stop in the street and stare, and their husbands' eyes would turn hard. I can still see his face in the darkness— generous mouth, hazel eyes—and the steadiness of his broad, capable hands. His flannel shirt smelled of the sun where Mama had hung it outside to dry, and the spice of fresh-cut pine hovered about him. In my earliest memories he is building, swinging a hammer so deftly that the fragrant pine receives a nail with only two blows. I see him mixing concrete in the space that would become the dining room of the Rushing Water Inn, pouring it between lumber guides, tending the fire all night so it would set up and not freeze. He built the rest of our buildings too, from the henhouse to the garage to the guest cabins. Daisy's house across the way was his work, and my grandparents' home farther up the road. He even built a guitar, and I can hear—or did I dream it?—the sound of him accompanying himself, not so much singing as mumbling "The Old Chisholm Trail" in the constricted range of a man who loved music but had no ear.

From the beginning, I had a case of the wanders, with a predilection for sticking forks into sockets or drinking fingernail polish. Consequently I spent much of the time when Mama was busy with the restaurant in my grandparents' care, or strapped in my high chair in the kitchen of the inn, watching the operation. My oldest sister, Rose Marie, was fifteen years old when I was born, and Mary was already eleven. Both of them were breathless to embark on adulthood, and by the time I was two Rose Marie would be gone, leaving behind almost no impression at all. But tall, pretty Mary is vivid to me still, down to the shapely ankles emerging from snowy bobby socks. Nothing eclipsed the flash of her dimples, a gift from my father, or her perfect teeth, full red lips, skin the color of cantaloupe. She was as luscious as those white cherries from the eastern slope, and I wanted to be just like her. Nothing made my heart soar like the sight of her swooping into the kitchen to stack four full plates along her arm. Out she'd go again, fearlessly balancing her load, her starched petticoats rustling under the fullness of her square-dance skirt, a jaunty little scarf tied around her lovely neck. I envied her freedom, her confidence, the way people brightened as if the very sun trailed her into a room.

My dominant impression of those earliest years was of being a complete nuisance. I desperately wanted to join the party everyone else seemed to be having, and I couldn't understand why I wasn't invited. It was as if my very existence was a continual surprise. Where the hell had I come from, and why? My father returned from the mines one winter night to find me alone in the house, the fires long out and a sheet of plywood nailed over the top of my crib to keep me penned in like an errant calf. I was cold, wet, and dirty, and I had pushed my empty baby bottle through the slats of the crib onto the floor. I still find myself paralyzed at the oddest times with the terror of being confined and abandoned.

I always fought sleep to the bitter end. One afternoon, Mama hauled me back to my parents' bedroom for a nap and got me down with the promise of a hunk of peppermint when I woke. I was playing with a bit of string when a shadow fell across the bed. I looked up, and there stood this dark, blue-eyed woman with a red bandanna cockeyed on her head, peering in through a gap in the curtain. Nobody heeded my screaming at the back of the house, assuming it was just my usual resistance to my nap. That image of a woman in the window is my first memory of Daisy, and it still makes my skin creep.

Later, when I was living with her, I would often look up and find Daisy gawking at me through the window. It was a habit she'd evidently picked up from the Klansmen down South, who had a streak of voyeur in them, and it was as revealing of her thinking as anything she did: It never occurred to her not to do something Klansmen would do, or to accord other black people more respect than history had generally allotted them. She just observed the advantage it afforded and included it in her inventory of skills.

The family enterprise thrived for a time, as my father had predicted. On top of cooking and washing the linens, my mother and Daisy doubled as hunting guides, taking the guests up the mountain to fish and to hunt for elk, deer, and bighorn sheep. In later years, Daisy claimed they were the state's first licensed female hunting guides, not to mention the first African American ones. She'd scream with laughter as she recalled the bed-wetting Texans with their big guns and loud mouths, who would get drunk and strut around with their balls clanking over the next day's hunt. In the morning, when it was time to hike up mountains covered with dense brush and trees, they'd collapse with altitude sickness. More than once, Daisy claimed, she or my mother had bagged their trophies for them, dressing the carcasses and bundling up the antlers for their clients to ship home and display over their mantelpieces. Meanwhile, Daisy and Mama revived strawberry and raspberry production in the area, selling the fruit and the preserves they made from it. They grew the produce and raised the chickens that were served at the inn. As Daisy liked to say, "We had the world by the tail on a downhill pull."

How wholeheartedly these freshly minted entrepreneurs were greeted is a matter of conjecture, but there is reason to think it was with relatively open arms, or at least a dose of western live-and-let-live. Strawberry Park was then a broad swath of income levels and political perspectives, the frontier mentality overlaid with the bohemian spirit of a seasonal arts community drawn to prestigious Perry Mansfield, a summer dance camp founded in 1913 by coal heiress Charlotte Perry and dancer Portia Mansfield. True, Daisy swore that forty-two miles from Steamboat, she'd encountered a sign that read, NIGGER, DON'T LET THE SUN SET ON YOU IN CRAIG, COLORADO. And there was a robust chapter of the Klan in Oak Creek, where my family got coal, but word was that they were more resentful of foreigners than of blacks, and by 1922 Steamboat Springs had passed a city ordinance prohibiting people from parading in masks.

However live-and-let-live the general attitude—and however pale a black man my father was—it was still the 1950s, and his impregnating a white woman, the wife of a local rancher, had serious implications. The two of them took off around the time I was two, and started a new family in California.

My father later claimed that Daisy's meddling had ruined his marriage to my mother, and Daisy in turn indicted him for a lack of ambition. The truth of their enmity probably wasn't understood by either of them, but one thing was clear. Daisy wanted vindication for the reversals of fortune she had suffered—for the humiliations and deprivations of the South, and her loss of grand wealth in Nebraska. And she wanted credit for saving the entire family in the process. Hers was the species of pernicious self-centeredness that made it permissible for others to do well only when she could be recognized as the source of their success. Whether any of this should excuse my father's abandonment of his family is another matter. I do know that his decision left my mother in circumstances from which she would never recover.

How long it took the business to completely founder, I am not certain. The seasonal nature of the hunting business, combined with intense competition from other guide businesses, soon made it necessary for my mother and Daisy to look for other means to supplement the family's income. Sometime between my third and fourth birthday, Mama's house and land went into foreclosure. Eventually our neighbors the Swigerts ended up with our home. I do not know the details of the transfer, but it marked the end of the time there when all things were possible.

One freezing winter day, there was a train wreck up at Rabbit Ears Pass. Mama packed newspaper under my coat to cut the cold, so that I rustled like a package as she lifted me into the wood-lined station wagon. The day was gray as ash, but to me, tucked between her and Mary—my sun and my moon—it couldn't have been brighter. Mama took the three miles into town as fast as she could, straddling the frozen ruts. I couldn't understand her hurry, but I was so happy to be with the two of them I couldn't stop kicking my legs. For once the invisible wall that always seemed to isolate me had disappeared.

I tried to stand up, to see over the dash. "Now listen here, Podley." Mama's breath hung in front of her mouth. "You have to sit still if you're going to go along. Otherwise, we'll just take you right back in the house."

I stopped breathing and froze as still as a pea in a pod.

Mary unwrapped a bright yellow stick of gum. "Want some Juicy Fruit?"

I could stare at Mary's dimples all day long. "What's *fyute?*" I asked, and was surprised when they both burst out laughing.

Mary's gloved fingers gripped my lips. "Furrrroooot," she said. "Say fuh-*root.*"

"Fee*yute,*" I replied.

They laughed again. Mary pursed her lips and blew a gray bubble. Absolute delight flowed up my body. The gods of my universe regarded me kindly. From time to time, Mama's coat sleeve brushed me as she turned the big wheel, dislodging layers of tobacco, woodsmoke, lipstick—and below all those, Mama's own particular smell.

The crunch of tires on new snow gave way to the smooth hum of well-graded road as we headed east toward Oak Creek. Time stretched out, long, elastic . . . boring. I chewed my gum, savoring the unfamiliar sweetness until it was gone. Then I swallowed it. "Can I have some more?"

Mary and Mama looked down at me sharply. The only thing worse than not getting any attention was getting the wrong kind and not knowing why.

"Rita Ann, did you swallow that gum?" Mama asked.

I tried to figure out what the big deal was. You chewed, you swallowed—what could be wrong with that?

Mary released her breath with a sharp huff. "She did. She ate the damn gum," she said over my head, looking at Mama. She looked down at me again, her lush lips pursed in disgust. "Hell no, you aren't getting any more."

Against all my will to contain them, tears popped into my eyes. Without meaning to, I'd ruined our sweet harmony, and our delightful, unexpected outing had deteriorated into a dreary ride into a featureless gray day. I always did the wrong thing at the wrong time, no matter how hard I tried.

We had just started up Rabbit Ears Pass when we came upon the locomotive. I had never seen anything so huge. Looming up out of the snow, it looked like an enormous wounded stallion. Behind it, a line of boxcars zigzagged haphazardly down the mountain on either side of the track, as if they had been kicked off. Mama pulled over to the side of the road and turned off the engine.

That's when we heard the squealing, so thick the air was saturated with it. I looked first at Mama, then at Mary. Finally Mama cleared her throat. "Tracks must have been iced over," she whispered. The wreck creaked and groaned as if it were a living thing, and I began to wail. "Rita Ann, hush up," Mary said, leaning closer to wipe condensation off the inside of the windshield. "Look, it derailed there at the turn. It must have happened in the last hour." I couldn't see what she was talking about, and I didn't want to. I tried to bury my head under Mama's arm, but she was in motion.

"Rita, you got to be a big girl now. We going to go up there and see what's what, and you have to be on the lookout."

"No," I shouted. But they were already buttoning their coats.

Then I saw the pretty oranges spilling down the mountain from the overturned boxcar, as if a huge crate had been upended. In the glare of the snow, the brilliant color seemed to vibrate. Beyond the oranges, another boxcar had jackknifed, so severely that it seemed to be in danger of rolling over at any moment. At first it looked as if the boxcar itself was bleeding, from a gash in its side. Then it became clear what the source of the gushing red mess was. It contained pens full of smashed pigs—bleeding, screaming, shitting, dying. Their snouts pressed between the slats, quivering, and the air brimmed with their terror. Some of the smaller pink ones had escaped and were floundering in the deep snow, their coarse, dingy hair flecked with blood. Most of the black and white ones were too big to squeeze through the opening, but one black bomb of a pig stood shivering in the wind, at least a foot taller than me. A large open gash on his hindquarters was pumping blood, and he dragged his foot behind him. He tried to run back around the boxcar but he kept sinking through the snow, until his scrotum was crusted with ice.

Soon other pickups pulled up. A crowd seemed to appear from nowhere, men with scarves wrapped across their mouths. "Mae, what do you think?" called a man.

At that point I climbed under the car seat for good, and that is all I remember of the day, except that we had a good deal of fresh bacon that winter. And that it was the last picture I would have of being with my mother in Steamboat Springs.

~~~

The Trailways bus was cresting a hill when I woke. Outside the window lay a bowl of lights, shining so brilliantly in the dark night that their reflection on Mama's glasses completely obscured her eyes.

"Mama, look," I said. "Hollywood be thy name." The night before, I had charmed her no end mispronouncing "hallowed" during my nightly recitation of the Our Father. But she didn't smile now. "That's just Denver," she said.

How could she not warm to this constellation lighting the underbelly of the sky? I drowsed against her arm, which still smelled of the hand cream she had put on that morning in Steamboat Springs. But she seemed far away. I sat up and leaned against the cold window. Outside, more cars than I had ever imagined possible raced along beside us like cattle down a chute. A low, strange roar rose from the highway, and there was a quickening in the air, although I couldn't tell whether it came from inside the bus or out. All around us, the other travelers were rustling their bags, snapping their pocketbooks, applying lipstick. Lighters rasped as cigarettes were lit, and the smell of smoke bloomed in air already stuffy with the scent of too many bodies.

The lights began to emerge in clumps, as the stands of pines that had lined the highway gave way to clusters of buildings. Now jerky neon zipped by on every side. I studied the signs and realized they were trying to tell us stories about what we could do or have. Cowboys lassoed blinking bulbs of red, lemon, green. Ladies in pale peach swim caps dived down toward the sidewalks or, less modestly, flaunted twitching hula skirts or fringe around their hips and tassels on their boobs. So many of the stories were about food—ham and eggs and pancakes offered by smiling white ladies with no troubles. How could Mama say "just Denver"? This place thrilled me with its extreme colors, almost as if the soda pop that boiled cold on your tongue had been trapped in those lovely neon tubes. I wondered if they were really big straws—if I could drink that light and it would taste of grape or cherry or orange pop.

At Mama's rooming house, however, it wasn't so cheery. We made our way up narrow curling stairs that smelled of mice droppings to a room jammed in the attic like a last-minute thought. Naked boards, rickety bent-pipe headboard, squeaky springs, a splat of a mattress that had taken just about all the sleepers it could manage. I lay there, marveling at how loud the city was, reverberating with the calls of machines—sirens, trains, brakes, tires hissing on pavement. I could even hear the refrigerator downstairs humming its cool song to itself. How could anyone sleep in this place called Denver? I couldn't remember ever having gone to sleep without the sound of the creek that ran by our house and gave the Rushing Water Inn its name. Nor had I ever slept without the whisper of trees over the house—cottonwoods, aspens, spruce, willows—or the brush of pine boughs. Denver felt naked, hard, without and within. Why had Mama come here?

But sleep I did at last, and woke to a birdless dawn, in a sky gorged with smells of burning rubber, a stink so thick my eyes watered. And in the afternoon, when Mama put me down for a nap and left for work, in a white uniform and a hairnet that looked like cobwebs, I slept again. In the evening, though, we went out. When we came to a set of those blinking red strawberry-pop tubes that spelled B-A-R, we ducked down the stairs to the door. Denver finally made sense. I had seen the neon; now I would drink the pop.

I was afraid at first, descending into a room where it was darker than the brightly lit night outside. Something heavy dwelled there, the atmosphere weighing on my lungs as if the air was too full—of sweat, ferment, booze, men's cologne and women's perfume, cigarette smoke and hair grease. There was something else too, dense and loaded, something that couldn't be breathed. Strong, squeezed music, nothing like the country western I heard on the radio at home, rippled out of a tutti-frutti box. Stools stood along a half wall like huge bolts in the floor. Everything gleamed the color of dried deer blood—the seats, the light, the men with their patent-leather hair, the dark-skinned women with their big red lips. I had never seen so many black people. When they laughed or talked, a disturbing energy seemed to be pushing them, as if underneath what they seemed to be saying lay something much more urgent. It made me a little wary. "Oh, just look at that head of hair," rasped one woman with a scratchy voice and a gold rim around her tooth. I drew back as far from her as I could, but she still got hold of one of my braids.

"She's shy," Mama said, tucking me inside a booth, the vinyl slick and cool against my legs and hands, then slid in next to me and sighed. I couldn't seem to get her to look at me. A lemon-colored lady with bright red fingernails brought a lovely little glass shaped like a tiny flowerpot on a stemmed pedestal small enough to fit my hand, but Mama wouldn't let me near it. When she lifted it to her lips, the golden liquid oiled its way up the side of the glass into her mouth, and she gasped. The aroma of rotten fruit bloomed in the air and overwhelmed me when I stood up to whisper in her ear. How could she drink something that smelled so vile? But then I saw the cherry in the bottom of the glass, redder than any cherry I had ever seen. I had to have it.

Mama leaned back against the booth like she wanted to sleep. The corners of her mouth began to soften. I jumped up and down on the Naugahyde for what seemed like a very long time. Then all of a sudden, she pulled the cherry from the glass, put it on a little napkin, pushed this treasure toward me.

"Go on. Take it."

It glistened a hot pink-red, like a cartoon of a cherry. When I put it in my mouth it exploded like candy, its flesh cool. Even the long reedy stem was sweet. In spite of the whiskey under-taste that burned my tongue and smoked my nostrils. I decided I liked this place called B-A-R. A lot.

When the lemon-colored waitress came by with another flowerpot glass, she brought me a little pink umbrella to play with. The men called Mama "Miss Mae Ella" and let me wear their Stetsons, which smelled of men's hair. By the time Mama and I went back out into the cold night, we were both warm inside, and nothing could hurt us.

I have no idea how long I stayed with Mama at the rooming house in Denver. Nothing was as it had been, and the moments I remember are haphazard and untethered, as if I were lost in a blizzard. Every day, it seemed, I was in a different house, some-times with people I knew, sometimes with strangers. By this time Uncle Billy had married and moved to Denver, and he and his wife, Helen, cared for me occasionally. The rest I can only interpolate from a mosaic of isolated images and impressions: The crop of moles splayed across the face of an older black woman. The scent of crayons as I lay on the floor, coloring. The wimple cutting cruelly into the pale skin of a nun. The mesmerizing texture of dust spinning in light shafts that penetrated the room and the odd silence of the sisters, those impossibly tall, draped, vaguely feminine beings wearing architectural constructions on their heads, as if they had donned gutters to collect rainwater, their faces and bodies mysteriously absent.

For some reason neither of us could later reconstruct, my sister Mary came to Denver and took me back to Steamboat just before Christmas that year, and my mother stayed on to work. Maybe she planned to follow with presents for us all, and couldn't manage a four-year-old as well. Maybe whatever arrangement she had made for my care had fallen through, or wasn't in operation so close to the holiday. To this day I wonder why I escaped her fate—or whether I might have saved her from it had I been with her, restless sleeper that I tended to be.

The next day a ferocious blizzard blew in from the Arctic to squat over the region, plunging temperatures below zero. All over Denver, people fired up their furnaces extra high. The heat must have comforted Mama after she slogged home from work that night, through knee-high drifts. I imagine her lying down on the lumpy striped mattress, out of habit leaving a space for me. I've been told that she had no blankets on that small cot, only shadows to cover her shoulders, a coat tossed over her legs. But it was good to rest her back, which had already endured a lifetime of bucking bales and lifting pots and babies and livestock. How hard it must have been for her, piloting someone else's stove after she had run her own restaurant, owned her own home and business, and had three beautiful daughters and a handsome husband. At least in Denver, she hadn't had to face his new wife's abandoned family at the feed store.

The deep snow muffled the sounds of engines, and the spring-loaded window shade blocked the draft, along with the pretty confections of frost frozen to the window in the moonlight. Still, it was hard to sleep, so close to the man in the next room, the walls made of cardboard, the restless city outside. The furnace-warmed air carried an unpleasant tarlike smell as it rushed up through the grate in the floor, nothing like the fragrant pine we burned for heat in Steamboat. But the fumes that came up with it, trapped by the poorly maintained chimney, did not betray themselves, and the building filled with carbon monoxide as my mother slept.

What was the last sound Mama heard? The man in the room next door sliding on his house shoes to shuffle down the hall to the toilet? The stroke of the match just before he sucked the flame into his last cigarette?

~~~

My aunt peered down at me through tears. "Honey." She seemed to be choking. "Rita. Sweetheart. Your mama's gone."

"When will she come back?" I asked. Mama had gone away before. She always came back.

Daisy tilted her head to one side and looked down at me. I had never known her to struggle so hard to be gentle, and this alarmed me more than anything she said.

"Honey, your mama done gone to be with the angels. You gonna be with me now, I guess."

Everything about her seemed to be trembling, dissolving, from her slumped shoulders to her run-down shoes. She smelled of iron and salt, as if she were sweating grief, clutching and tugging at a dish towel she had been using to dry dishes when the phone rang. My eyes were level with those work-roughened hands, grime under the nails. I could not stop staring at her left hand's ring finger, which had been lopped off somehow, leaving a pale, mushroom-colored nub.

Did she mean for a long time? All night? Why would Mama run off with the angels and leave me with this old woman with narrow lips like hers, but nothing else familiar? Why did they call her Daisy, I wondered. With those strange rheumy eyes and light-blue rings around the irises, she was nothing like the daisies that grew in our yard.

Sometimes, I knew, I had dreams, and when Mama woke me up I discovered they weren't real. Like in the movies, when Tarzan and the lady were there and then the lights came on and only a blue velvet curtain remained.

That nub-fingered hand was resting on a table that held a jumble of tools and food, papers and bobby pins, pens and sugar and crumpled geranium leaves. Daisy clutched it now as she spoke again. "Rita, Mae's passed away." She gave herself over to a silent openmouthed spasm, clenching her fists, doubling over. "My baby sister is dead. Oh God, how can that be?"

If I closed my eyes, maybe Daisy would disappear and Mama would come back. But when I opened them again, there she was still. I tried another tack.

"If I was dead, could I be with Mama?"

"No, honey. God's called her home."

I couldn't understand her. Home was just across the road, down past the creek. I could see it out the window. Now that I was four and a half, I could even walk to it by myself if Lady would just watch out for snakes. Where was Lady?

"Reeter Ann, the Lord giveth and he taketh. When your time's up, you got to go."

I didn't like the way my name sounded in Daisy's mouth. It

felt unsafe. If I could just get back to Mama's house, the natural order of things would be restored. Everybody had a mother. Even Daisy still had Gama.

"You got to give up being babyfied now. Thank Jesus you ain't laying in a casket too."

I began to float out of my body, high up into a corner of the room where I could watch the two of us from a safer distance.

"Do you hear me talking to you?"

I had gotten lost somewhere in the motion of Daisy's mouth, watching the way her false teeth slid up and down her gums. So, no, I hadn't heard her. In fact, I had forgotten the beginning of each sentence before she got to the end. But already I knew better than to say so.

And then it was a different day. How could the sun still dare to show itself? How could the snow keep falling, frosting the pines in the yard like gingerbread? Still no Lady, though. I hadn't seen her since I got back from Denver. "Daisy, did Lady go to be with the angels too?" She turned away from me as if I had not spoken.

The aching did not abate, not in the slightest. Morning, afternoon, outside, in bed, my life was a wreck that spiralled on and on. It seemed incomprehensible that I was still required to pray— *If I should die before I wake*. Every night I kept watch down the road. Mama had been real once, had really passed over that horizon, so why couldn't she cross back the same way she had gone? If God wanted to, he could make it happen.

Maybe if I was really good, prayed hard enough, did my chores. Fed the stinking hens with their retinue of stupid chicks. But in spite of everything I did, Mama did not come back. And worse, the memory of her face began to fade. The harder I chased it, the faster it receded, until it was as if she had been a dream that refused to be recalled but colored every waking moment.

Time passed, indifferent. The snow melted and spring burst upon us, with its urgent dandelions, white-faced Hereford calves, and frisky foals. Mary raced off to school like a palomino with a new silver bridle. A new puppy was tied outside to the willow tree, but when I passed on my way to Gama's house, I ignored him.

Gama put a pan on the oven door so I could wash dishes too, gave me a lump of dough to roll out alongside her when she made fried peach pies, saturated me with quiet kindness. Now when I walked up to her house, the new puppy they had named Jiggs stood up on his hind legs at the end of his chain, yipping and begging me to come over and play. Daisy and Mary told me he was one of Lady's puppies, but they still didn't answer when I asked where Lady or the rest of her puppies had gone. Jiggs looked nothing like her. Where she had been black and sleek-headed, glossy as coal and light on her toes, he was chunky and square as a brick of cheese, more like a little bear cub than a dog. He had huge clumsy paws, fur the color of creek gravel tipped with black, and a light creamy brown undercoat. I hated him for not being his mother.

He put up with me holding a piece of venison just beyond his reach and then popping it into my own mouth. He let me tie a rope around his neck so that he could pull my sled down the road like Rin Tin Tin, choking all the while. Gradually, his re-lentless adoration won me over, and one night when I was sit-ting outside with him in the cold night air, I snuggled up against his warmth and it occurred to me that both our mothers were gone, and we might as well look after each other.

One night Mary and Daisy got the idea of taking me to see *Bambi*. The movie enchanted me at first, full of images I could collect and take home to savor later. In Bambi-land, none of the colors were dull and a chorus sang all the day long and a hand-some dad stood by, deep-chested and vigilant. How did you get to live in that forest where nobody got pecked? Why couldn't I go there?

And then the rifle shot. I had heard that sound before, and I knew it had no place in a movie about a baby deer. I watched as Bambi ran to the safe place, and then held my breath as he turned around. It had been the same with me. I had gotten away, but Mama had been killed. Mama, a hunter herself, who could put a bullet through the heart of an elk a quarter mile away.

Then Bambi's father—the buck with the big rack my mother would have taken as a trophy—spoke through the curtain of snow.

"Your mother isn't coming back."

I screamed so loud they had to carry me out of the theater. And I knew then, at last, that Mama was dead and she wasn't coming back.

3 ~ The State of Water

A team of draft horses working a Steamboat ranch

~~~~~~ Colorado—the rhythm of the name sounded like water flowing downstream. Daisy always said, "If you ever get lost, follow a creek, a river, or a ditch. Sooner or later, water will lead you to safety." Of all the problems that beset me and my family in that state, we were never denied the blessing of water in all its forms—ice to skate on, flush creeks to fish, mist on the meadows in the early morning, pelting April rain. There were even steaming mineral springs, liquid and gas all at once,

that reminded early white residents of steamboats back home, giving Steamboat Springs its name. But none of these forms compared with Steamboat's bone-dry champagne powder snow. Some years we'd get twelve feet by Christmas. A skier could sail down Mount Werner, no drag on the skis, thirty-six hundred feet to the bottom.

Away from the tourist scene, in Strawberry Park, the deep, diamond-laden drifts swallowed sound. At dusk I loved to walk up the road with Jiggs toward Gama and Gampa's house, past the last cabin, all the way up to the barn at the edge of the property that had been Mama's and had now been annexed by the Swigerts. In the stillness, there was nothing but the sound of my footfalls. The snow absorbed even the padding of Jiggs's paws. Sometimes the stillness felt so dense the winter seemed to be listening to itself.

The black scars on the pale aspen trunks appeared to be eyes watching. I could see the lacy prints of weasel feet all the way to the creek. Across the buried pasture, the diagonal zigzag of a rabbit—one that got away. Daisy had showed me how to snare snowshoe hares the way Robert Ball Anderson had taught her, with some string looped over a strand of fence wire. She told me that without those rabbits my parents and sisters wouldn't have survived that first Colorado winter, when the snowdrifts blocked the door and they had to crawl out the window. It astonished me to remember that even in the deepest cold the land was full, bursting with life. Under the drifts were all those dreaming, hibernating bulbs and seeds. Beaver families kept house inside their dams, and trout lay in the creek. I wondered if the trout slept, and if they did, whether they dreamed of summer.

I sang to myself on the way back down the road. A painterly trail of smoke streamed up from the chimney above Daisy's house. I still couldn't believe I lived there now. A mighty blue spruce surveyed the yard, and a couple of snowbirds hopped among its upper branches. It still felt as if I were dropping in to borrow a cup of sugar for someone—Gama, maybe. Not Mama, whose face I could no longer remember, though her absence filled every corner inside me. Finally I'd realized that when people "passed away," that meant they were dead. I still couldn't help looking down the road at the sound of every engine, though, hoping she would come. Maybe one of those angels that had taken her might fly her back. We certainly had enough other winged creatures about.

Daisy had left the backside of the house unpainted, because she thought nobody saw it driving down the road from town. But to anyone coming back down the other way it was visible, along with the series of shantytown sheds oozing with donations from people Daisy had worked for over the years. It seemed to me the discards of half the town had found their way into the boxes, crates, bushel baskets, bins, and tins stashed back there, where they fermented in their own history. Broken-zippered skirts that smelled of men's cologne, and eggshell linen pants stained with blood. Old sheets worn down to threads. The stained embroidered doilies of the lately departed. Daisy couldn't say no to any of it: "I don't believe in throwin nothin away." But this did not mean she would actually get around to using any of it.

Not all of the sheds were devoted to this junk, however. Sometimes they housed snowy waddling ducks, or speckled guinea hens who always managed to look alarmed. At the end, next to the outhouse, the noisy turkeys resided. Although their dark circular plumes could be magnificent, they were so stupid they could get lost between the yard and the roost. "All their good sense been bred out of 'em since they come to be about Thanksgiving," Daisy said. There were also four Rhode Island

Red setting hens, a lone white peacock with an attitude that would peel paint, and two new additions, a pair of old geese I called Amos and Bess.

By evening the shed stunk of poultry shit, but after I raked it out and clipped the twine on a new bale of hay, the scent of languid August was in the air. I liked to listen to the birds clucking and fluffing their clean new bedding, settling in for the night. I could hardly bring myself to leave them.

"What in the Sam Hill can you be doing out there in the henhouse every night?" Daisy would demand, pounding venison with a mallet. To her, animals were livestock, not pets. Horses were stupid and easy to spook, and she couldn't understand anybody riding for pleasure. If she knew I'd named the geese, she might have slaughtered them on the spot, just to teach me a lesson.

Ira had brought them. He and Daisy had had a contentious relationship about water rights. Water was the thing in the Rockies. It also hadn't helped when he'd caught Jiggs chasing his blue-ribbon Hereford bull. Ira promised he'd shoot that dog before he'd let him run the fat off his beef. But neighbors we were, and neighbors we stayed, and one late fall day Ira's pickup had clattered its way up our road. He began this visit just the way he'd begun most of his conversations with Daisy: "Thought maybe you could use these." And then he swung down a crate that honked and squawked in indignation.

The crate turned out to contain an elderly goose and gander. Daisy went to get the hatchet and whetstone while I studied them. It occurred to me that they'd arrived on Daisy's doorstep much the way I had. Even incarcerated, they retained considerable dignity. The gander's plumage was sedate as the suit of a banker—layered shades of silver, oyster gray, and charcoal. The feathers on his neck clumped together, in fine honeycomb patterns that rippled when he moved his head. His mate was surprisingly large for a female. Her feathers were so white she seemed to be covered in layers of lily petals. She would be invisible in snow were it not for her blazing orange beak and feet.

Daisy laid the whetstone next to the crate and put the ax with its new handle in a bucket of water to swell.

"Daisy, these geese will be tough as rawhide," I said when she returned, trying to sound like a wrangler.

"Well that's a fact," she said. "Ira wouldn't have given me nothing he could have gotten some good from." As she studied the geese, I stayed quiet. It wouldn't do to pressure her.

"If we get these two through the winter, we can get a load of goslings from the hatchery come spring. Most times the old ones'll go to settin just like those babies was their own. We might just end up with a flock of geese to sell."

Money spent on feed meant money not spent on winter coal. Daisy pulled a square carpenter pencil from the bib of her overalls. "I need to figure whether it's worth it to feed this old pair all winter long." I jumped to bring her a paring knife to sharpen the pencil along with one of my Big Chief tablets to write on. She whittled down the lead and began to do the calculations, in her old-time scrawl. "All right, if we keep them in the shed and they don't run around, I can afford to see if they'll set on the hatchery goslings. If I can sell fifteen for the Christmas season, we won't have to worry about coal next winter."

~~~

By the end of March, we had used up almost all the choice cuts of elk and venison, even the freezer-burned charity Christmas turkey, which had been dry as dirt. About all that remained were the tough shoulder roasts that we had passed by the year before and that tasted like the bottom of the freezer no matter how much garlic Daisy studded them with, and chicken feet Daisy boiled with the toenails on. The sight of the yellow skin shedding in the scalding water was a sure sign we were in the part of Colorado winter tourists never saw, the part that seemed to drag on forever. It wasn't bright and shiny any longer, just dead cold.

Mary had taken a job at a radio station in Craig, forty-two miles away, that required her to leave the house before dawn and returned her well past dark. Daisy cleaned houses while I was in school, and some days she'd leave me at the library while she cleaned the church in the afternoons. She'd come to pick me up smelling of sweat and Clorox, and when we got home she'd build a fire so hot the chimney glowed, toss a package of deer ribs in the Dutch oven, and boil them together with potatoes and frozen dandelion and mustard greens until the mixture turned gray as the seat of a Confederate's pants—which I knew about because we'd gone to see *Gone with the Wind*. It seemed that Daisy had come home determined to use the last of her energy on punishing the food until it surrendered.

Nights like these Daisy ate without talking. Eyes glazed, she stuffed the food in with both hands until her mouth bulged and her pretty lips, with their two symmetrical moles, were slick with deer fat. She didn't so much eat as feed, sinking into a trance. When she was like this—hunched over the plate, grease on her chin, sopping her cornbread in the pot likker, she disgusted me—and frightened me. But God help me if she looked up, caught me staring judgment at her. Then the glaze left her eyes altogether.

"What you studyin, starin at me like a bay steer?" she snapped one night. "Thought I done told you to eat your supper."

My eyes dropped to my plate, too late. She didn't like to be watched when she went to that private place.

"You may as well get started, because you gon' clean that plate." When her voice went that soft, I was in bad trouble.

I didn't say anything. In the tepid light from the pale lightbulb, I pushed the defeated greens toward the potatoes. I pinged my fork against the jagged venison rib, which was cracked and oozing marrow. I picked it up and took a bite. Most times, Daisy was good with seasoning, but this was like chewing elastic. I tried to breathe through my mouth so I wouldn't have to smell it. I swallowed a chunk whole. If I could just get it to him, Jiggsie would help me with the rest of the deer meat.

But the venison wasn't the biggest problem—I could pour on the salt and get enough down to make a good showing. I simply could not pick up the powdered milk, a jelly glass full, the color of maggots. The lumps floating at the top would tickle my lips and make me want to puke. Daisy still resented the welfare woman who had criticized her because we had a Deep-freeze but no refrigerator to keep the fresh milk that would help me grow long, strong bones.

Now Daisy waved a long, curved rib at me that had been picked clean of everything except the gristle.

"Eat up," she said, her mouth full.

Panic filled me. I prayed a single, strangled Our Father to the sad crucifix hanging above the window, with its fine dusting of coal ash.

"Daisy, my stomach hurts. I'm not hungry."

"Did I ax you how you felt? Or if you was hongry? I don't believe I did."

The plate of food lay in front of me like a sentence. Outside, the dense air concentrated the silence.

"Answer me. Did I ax you if you was hongry or did I tell you to eat?"

"No and yes." Where had that edge of anger in my voice come from?

Daisy's eyes locked on me now, absolutely clear and bright. "I done cleaned two houses today. Woman didn't have the money. Say she'll pay me tomorrow. So I make eight dollars, come home to cook for you, and you too good to eat it?" This was going to be bad.

And then the phone rang, startling us both. Jesus, Mary, Geronimo, the phone rang, the phone rang, the phone rang. We had a party line. Four rings was our signal. Daisy pushed back from the table, fixing me with one last look. "Finish up now, Reeter Ann. I mean it." She found a rag to wipe her hands on and picked up the receiver.

"Hello?"

Daisy had a special voice for everything. One for chatting up white people. One for babying the geese. And, Southern woman that she was, one for charming men of any color. On the phone, she used a very loud register, as if she couldn't quite trust she could be heard by someone she couldn't see.

"Ernest? Is that you? You don't say." Ernest was the younger of the two brothers who were still living. I had no idea where he lived, but he had appeared at least every six months for as long as I could remember and disappeared just as suddenly, trailing a plume of Tokay fumes after him. He never answered my questions, and I knew he didn't have much use for me. I didn't like him much either, although he always brought great food. Once he'd brought us a piglet, which in no time at all grew to more than a hundred pounds, when to my relief Daisy sold it. But Ernest remained her favorite and I knew she'd be on with him a while, suspending her usual aversion to using the phone for much besides making the briefest of plans, for fear that all of Strawberry Park was listening in.

The phone stood on a little table by the door, about ten feet from the dinner table. Next to the door were windows Daisy had nailed an extra board across to make room for her infernal collection of geraniums. They smelled strange to me, but she plied them with coffee grounds and eggshells and caressed their velvety leaves as if they were flowering lambs. As soon I saw her begin to snap off desiccated stems, I made my move. I stuffed two ribs down the front of my pants.

"You reckon? Well, ain't that a play-pretty."

I darted past her and out the door. Jiggs took the bones back under the porch to bury for later.

When I came back inside, Daisy was still absorbed in crumbling dried blooms into the clay saucers. "You'll be here day after tomorrow? All right then, we'll pick you up." She hung up and turned around.

"Ernest fixin to come. Must be he won something at the dogs."

Ernest loved to gamble, especially on the greyhounds. He followed the circuit all across the Southwest. He loved small dogs too, but in a darker way. Each time I saw him he seemed to have a different one. They had a habit of disappearing.

~~~

"**Get up.**" I turned over just as the covers were yanked off my body. Behind the round cold light of the flashlight, Daisy loomed at the foot of the bed. "Get up right now. You fixin to get a whuppin."

The weak beam sliced through the darkness, revealing the belt in her hand.

"I'm sorry," I whispered, curling myself into a ball.

"Get up, heifer." The belt whistled down across the back of my thigh. I screamed as it bit into me, and jumped out of bed and ran into the wall. "Get to the table." I could barely see her face behind the flashlight.

I struggled to find a way around the belt, but couldn't. I tried not to cry. Whimpering fed Daisy's fury like kerosene.

She grabbed my arm and yanked me back to the table, where the flashlight illuminated the powdered milk I'd left at supper. How could I have forgotten that?

"Drink it," she said.

I put the glass to my mouth. The milk had clabbered alarmingly, like spoiled custard, not like cow's milk at all. I held my breath and took a sip. It gagged me. Daisy stood by the red leather chair in front of the washing machine. The Deepfreeze hummed behind her.

"You reckon I don't know you been slippin my expensive food to the dog?" She popped the belt.

That familiar floaty feeling had come over me. I felt as if I had slipped above myself and hung there, watching.

"I catch you feedin your supper to the damn dog again, I'll tan your hide so bad you'll wish you was dead. Now drink that milk."

I took a big mouthful and tried to swallow, but my throat refused to take it. I threw it right back up into the glass.

"Puke it up one more time and by God, you'll lick it off the floor."

And then I saw something shift inside her. She seemed to be looking past me, struggling for breath. She opened her mouth as if to scream, but no sound came. It went on forever, as if she was suffocating. Finally she sobbed, tears bursting from her eyes, the belt limply snaking from her hand to the floor. The crying upset me more than her anger. Then she found her voice.

"Mama had to wet-nurse every one of that landowner's kids, come home with no milk in her titties for her own. Worked all day, cooked them dinner and supper too, then washed up, fought off ol' man Patterson, walked home in the dark to climb up under John Wesley and give him some. And me there the whole time with all seven kids, and I'm not but fourteen."

Who was she talking to? Not me, I knew that.

"I worked there too. Seen how it worked. Had to serve at dinner. Some had clabbered, some had sweet. Other'n had buttermilk and still t'other drank straight cream. The woman liked to make out like she was bein real nice, had me pour all the milk glasses together in a pail to take home like I'm fixin to slop some hogs. And you know white folks is nasty. Won't wash they face and hands to save they soul. Just sit down to the table nasty. It's because of them our folks come up with TB and whoopin cough and smallpox and mental."

She had been pacing the floor, back and forth between the table and the Deepfreeze, but now she got quiet. Her shoulders stopped heaving. She turned and saw me again. "Don't never take nobody else's child to raise. You hear me?"

She wiped her nose on her sleeve.

"You puke up that milk ever again and I will lock you in the cellar. I mean this. You gon' be strong when I git through with you, Rita. Strong, and a credit to your mama."

~ ~

At the tail end of April, Daisy woke before first light and made a huge breakfast—the last of the elk steak and fried eggs and canned peaches and big hunks of bread toasted on top of the stove with a little webbed wire. Frost still speckled the windows when she called me to the table. Chewing a piece of meat, Daisy pulled her fishing pole out from behind the buffet and began to examine the line. "What ya think, kid? Twenty-pound test?" She pulled it taut between her fists, nodding when it held, then turned to me, her eyes bright as a girl's. "I'm going to throw together a couple of elk sandvidges. Get on your overshoes."

I didn't much care for fishing, but I knew it would put Daisy in a good mood for a week, whether she caught anything or not. Nothing got her going like the prospect of waiting, sometimes for hours, over a fishing pole. When she finally got a bite, she'd go berserk. "Great God, I got a strike! And he's a big'un." The only thing that made Daisy happier than fishing was plotting how best to fish. Still, I didn't see us going anywhere with ten feet of snow on the slopes.

Before I could say anything, Daisy went out to the shed. She came back with her wicker fishing creel stuffed with some of the fragrant, pale green hay we used for the yard fowl's bedding. She had a coffee can of worms too. Worms were treasured livestock to her, and she kept her plot secret, supplying it with potato and fruit peelings, watering and tending it, even moving it when she suspected a neighbor might have seen its location.

She went into the kitchen and buttered thick slabs of bread for elk sandwiches, which she wrapped in reused aluminum foil. Then she began to wrap me too, in layers of coats and sweaters. "Now put a wool scarf on your head."

"I can dress myself," I said, but she kept on tugging a strip of wool cut from an old sweater around my throat as if I hadn't spoken at all.

"Now do your business before we get up there," she said, meaning use the outhouse.

Just when we were packed into Daisy's World War II–era Jeep and she had her key in the ignition, she stopped. "Oooh, wait a minute. Nearly forgot the gold." She bounded back up the stairs into the house and came out with a wedge of welfare cheese.

"For us?" I asked.

"For the rainbows," she replied.

On the other side of Rabbit Ears Pass, the lake looked like just another vast field of snow. Against the horizon, the two tall pointed slabs of granite that gave the pass its name seemed painted on the sky. There was no way to tell where the shore ended and the water began, except that beavers had built a wide dam at the far side, where the creek fed the lake. There were no other signs of animals except for a few coyote tracks, nor were there any birds. Nothing else but the occasional crackling of the ice, and the intermittent gusts of wind. I thought to myself, *This is going to be boring as dirt.*

Daisy strapped on her snowshoes and headed for the center of the lake. I was light enough to walk on top of the crust. Before long, sweat was streaming off her forehead and I could smell the tang of the elk sandwiches heating up as the creel bounced against her thigh.

"We got to be careful not to get too far out," Daisy said, meandering from one seemingly identical spot to the next, listening—to what, I could not say, because I certainly didn't hear anything. Finally, she found the spot she was looking for and spread a tarp for me, covering it with the elk-hide rug that had been mine since I was an infant. She pulled a hatchet out of her creel and set about widening a small hole in the ice.

"Will I have to fish too?" I asked, careful not to whine. I dreaded the prospect of numb frozen fingers trying to thread a squirming worm onto a hook.

"Later, maybe," she said. "But right now, you be still. These fish heard every step we took getting out here." I looked at the expanse around me—nothing except for a small herd of clouds gathered at the far end of the horizon as if they were watching us. I settled back into the elk hide. As the sun climbed higher, it began to smell. I dozed off. Then I dreamed that I was falling, and jerked myself awake to see Daisy perched on her folding stool, motionless and patient over the hole. I have no idea how long she stayed like that, but it must have been hours, because, when I woke to her shout, the sun was lower in the sky again.

"Oh goddamn!"

I sat up. The line jerked this way and that, so hard that it looked as if it was going to slip from her hand. Daisy pulled the line steadily until a little silver fish face bobbed above the surface of the ice.

The fish was almost too big to squeeze through the hole in the ice, so Daisy got down on her hands and knees and, reaching down into the slushy water, put her finger into the fish's mouth and yanked. It popped out as if it were being born, flexed its fantail, and gasped in the unlikely air. Daisy eased the hook out of the trout's gills, put her thumb in its mouth, and bent the head back.

"Daisy, you're going to hurt the fish."

She laughed, low in the back of her throat, and looked at me as if she was sizing me up for the first time.

"That's the kindest thing you can do to him. Kill him fast so he don't suffer."

She could feel me watching her, struggling with that conundrum.

"Go on back to sleep."

Of course, I couldn't. Over and over I'd be drifting off when I'd hear the snap of another neck. I sat up. The stand of clouds that had been at the far rim of the sky seemed to have gathered mass, and one big cloud hurtled in front of the sun, turning the snow-covered surface of the lake steel-gray.

I couldn't feel my heels. They'd slid off the edge of the tarp when I'd fallen asleep and now when I tried to move them they were frozen. I was cold on the bottom, hot on top, shivering and sweating.

"Dais——"

"Hush up," she hissed, before I even got her name out. "You think these trout can't hear? Damn it all to hell." Her irritation stung, especially after she'd been nice to me all day.

I did my best to be quiet, but the tears wouldn't be held back. When I risked a soft sniffle, she tilted her head to one side. I knew she was wrestling, trying to decide what to do. Finally, she sighed, stood, and pulled the line up. "All right, we ain't getting nothing else today. I reckon we got a pretty good mess of fish here anyhow." She blew on her fingers and then stuffed them under her armpits. She stomped her feet. "How did it get so late?" She looked around as if she was coming to. "Here," she said, unwrapping a sandwich. The gray elk suet had hardened and the bread had softened, but it didn't bother Daisy. Some displaced persons had stayed with Mama before I came along, and Daisy had been very impressed with how they smeared lard on slabs of black bread. She liked to adopt the ways of people she thought were shrewd.

"Ain't they pretty," Daisy said, looking at the mess of fish that lay on the ice like large silver leaves. At home in the creek their colors didn't show, but now I could see that they were freckled, with a splice of rose up the middle. Again a shadow of cloud fell over the lake, rendering the fish a ghostly gray. Daisy was tucking them carefully into the creel when we heard the softest thump in the distance—like the shot of a .22 or maybe a tree cracking in the cold.

Daisy stopped breathing and stood up listening like a bear. When nothing else happened, she began to break down her pole into parts. She had just folded up her stool when we heard it again, a muffled clap like thunder in the distance.

"Rita Ann," Daisy said. "Let's go." There was a softness to her voice that worried me much more than when she spoke sharply. "No, don't bother to wrap up that bread. Let's get out of here right quick." She hauled me up and buckled my feet tighter in my overshoes. I cried out at the pain. How would I be able to run on my frozen heels, tangled as I was in layers of coats, sweaters, and scarves?

"Hustle, honey. This lake's a-fixin to go." I couldn't quite understand what she meant. Go where? But she shot me a look that silenced me. Her snowshoes strapped on, she took off, dragging me by the hand.

I couldn't keep up with her. Then I fell. For once she didn't yell or spank me. She just bent down and scooped me up under one arm like a sack of feed, the strap of the creel over her other shoulder, and kept moving as fast as she could in the soft snow. Finally, she dropped the stool to free up her hands and lumbered on.

Suddenly she veered off to the left, stopped, and quickly retraced our tracks, still with that deadly sober quiet about her that made me want to bury my head under her shirt. Then the thunder really started, rippling waves of it, as unseen cracks collected under our very feet wherever we stepped, the weight of our bodies drawing the fissures to us and radiating them away from us at the same time. We could hear it and feel it but we couldn't see it. Eerily, the surface of the lake remained serene as a snow-covered field.

Then the sound of the cracking stopped. Was it over? At about fifty feet from shore, Daisy seemed to feel we were on higher ground, but she must have miscalculated, because a shelf of ice gave way and suddenly we found ourselves in icy water up to my waist. It was as if the lake had reached up and grabbed us.

"*Whhhoooooooeeeee,*" Daisy yelled, and held on to me so tightly I thought I was going to break in half. "Bless Jesus, but that's some cold-ass water," she said as she wallowed me up on shore through the icy mud, creel held high.

I couldn't stop screaming, but Daisy started to laugh, thrashing through the icy mud. "Lordy good God, that was a close one." The lid of the creel had come unfastened. "Damn shame if after all that, we dropped them trout." She peered inside, then, apparently satisfied, turned back to me. "Rita Ann Williams, let's get the hell off this pass."

At the Jeep, she made me pull off my frozen jeans. My teeth were chattering so hard I thought they would break. She rubbed my legs dry with a towel like she was scrubbing down a horse. Then she wrapped me in the thick old quilts she always carried in the back.

"All right, kid. It ain't nothing but some cold water. Quit squealing like a stuck pig. "

She started the motor and then the heater, which normally was way too hot for me but now felt like God's purest mercy. I huddled as close to it as I could. A thick, gauzy snow had begun to fall in the graying light.

"*Hoooeeey.* That was a thing or two, now wasn't it?" Daisy looked quite satisfied.

She put the Jeep in first, and I took one last look at the lake. The surface had buckled and cracked all the way across, and a dark split had opened like a seam along the path we had been running. Daisy pulled out the throttle and carefully guided the Jeep astraddle the ruts down the hill.

"Winter fishing is the top. Look at what we got, nary a one of them under a pound. Besides, fishing is the best damn fun there is."

~~~

Winter ground its way down to a slushy, mushy spring, and at last it was time to go meet the train bearing the goslings, which arrived at a chilly 4:00 A.M. When the crates were unloaded from the freight car, we covered them with blankets and loaded them into the Jeep. At the sound of the engine, they squeaked in fear. Daisy began to sing to them, and they settled. All the way home, as we passed diamond frost clinging to the timid clumps of new pasture, each bump and hole in the muddy road set them off anew, and each time they calmed at the sound of Daisy's voice.

At home, we opened the lid as gently as we could. Inside, sixteen mounds of warm, olive-green down huddled together in the middle of the box. For an instant, they were still, as if stunned. Then they stampeded to the far corners, peeping loudly, except for one. It had gotten turned over onto its back during the journey and shoved into a corner, and now it was alive but lethargic. "Well, we got one down," Daisy said, regarding it with an experienced eye. "We'll see if she makes it through the night."

She reached in and lifted out the others, one panicked little gosling at a time. She checked them thoroughly, running her large, dark fingers over their bodies, stretching out their legs and wings, feeling their backs and breasts. "Oh, my little baby choo-choos," she sang. While she worked I took one of the babies out of the box and tucked it under my shirt, where it became instantly peaceful. Its little webbed toes were cool against my stomach, its belly warm.

All the goslings were robust except the one. We put a heat lamp in the middle of the floor by the washing machine and got up every three hours for a week, to make certain they did not turn over on their backs and die. Finally, even the weakest was pushing for her own feed and stretching her little wings. "We'll move them tonight," Daisy said. "It's starting to smell like a henhouse in here." In the middle of the night, when Amos and Bess were asleep, we took the babies out to the goose house and slipped them under their wings. When they woke in the morning, it was as if goslings had hatched from their own eggs.

~~~

Now spring was truly upon us, and the basement walls began to seep. By the beginning of May, the melt intensified until we could hear water dripping down there, even from upstairs. In the morning the puddles on the concrete floor were so deep that I had to walk on boards to go fetch a jar of peaches for breakfast.

On the slopes, the snowpack appeared solid still, but underneath it was melting. The ground was a sloppy mess, unstable, slippery with loose leaves. The creek by the house, which had been muffled all winter, began to rise, nearly overflowing its banks.

Cows finally let out to pasture came round in the evening, their faces caked with mud kicked up by the rest of the herd. They looked dismayed somehow, their vacant eyes confused behind their long, pale, feminine lashes. Daisy drained the water from the tub of ashes she'd been stockpiling to make lye, and hauled out the canisters of bacon grease she'd been saving all winter to make her brown soap. She made me stay in the house when she made the lye, though, away from the fumes and the roaring fire in the yard.

The cranes came back from the south, as well as the Canada geese in their familiar chevron, slicing through the sky with strong, steady strokes. The big lead bird broke the wind for the others until he tired and dropped back, letting another take his place. High overhead, their honking aroused our geese, agitating some vestigial urge to migrate too. Pussy willows popped up along the creek. At midday now came the first surge of heat and the prospect of warm hands and feet again.

Slim McCormack brought his paint mare Magpie to our upper pasture, along with her new foal, Thistle. He rented this land and the cabin on it for the summer. I liked it when he came down to visit Daisy, his blue eyes clear and patient. He listened to whatever she had to say. He was a carpenter now, working for himself, but for years he had been a working cowboy, and he still had the bowed legs of a man more often on horseback than on foot.

In the meadow, brand-new dog-sized calves appeared, with red bodies and white faces. And suddenly there was grass, all in a rush, and the aspens blurted out new leaves. Daisy was intent on weeding now, literally from dawn to long past dark. She worked the shovel while I got down on my hands and knees and sifted the soil to find every last hair root, so we didn't have strong weeds taking the strength of the strawberries later. She watched to see how low the temperature dropped at night. Time to plant the leaf crops—cabbage, lettuce, spinach, greens. Like the rest of the world, I came to in the spring, awakening just as the sky burst in—a rolling herd of cumulus, an avalanche of cloud. I couldn't help but feel the giddiness of Slim McCormack's new foal after it survived that first storm, any more than I could fail to respond to Jiggsie with his elbows on the ground, butt in the air, ready to play.

Jiggs was losing his winter fur now in thick, matted hanks. Sometimes I could get Daisy to let him off his chain and go walking with me, but he almost never came when I called him, returning hours later with a face full of porcupine quills. He couldn't seem to remember that they weren't so much running away as luring him close so they could swat him. The musk of skunk was in the air—they were having kits too—and weasels had shed their snow coats at last.

Now the creek boiled with mica stirred up by the current, golden brown like Jiggs's eyes. Daisy, coming upon a school of rainbow minnows tucked behind a still eddy, was overcome with sudden tenderness: "Ain't they sweet?" But that didn't stop her from catching barely legal "brookies," as she called them, for breakfast. They were delicious breaded in cornmeal and fried crisp. While she fished, I perched on the creek banks laden with low, flowing grass that trailed in the water like green hair and gathered panes of ice to eat, the water still so cold that within a matter of minutes my hands and feet felt hot, then numb.

Colorado spring, bigger than God, continued to erupt, in a geyser of explosions, as if that forty-five-day growing season made the green beans catapult up their guides in a fortnight. Soon the peonies bloomed, like fistfuls of pale pink feathers. And in the field outside our door, tender young greens gleamed under a Milky Way so bright the geese had to tuck their heads under their wings to sleep.

~~~

We were on the eastern side of the house—one of the painted sides—planting potatoes between the house and the creek. The blue spruce was fragrant with new growth, pale green shoots at the tips of its fronds. From the shed, Daisy brought the shovel, the hand spade, and some carefully aged chicken manure. She went to the cellar and carried up the last of the potatoes. With a curved pruning knife she cut them in half, then in half again, careful to check that each quarter had a viable eye.

"Reeter, run get the can of turpentine so these bastards don't eat us up," she said, swatting at mosquitoes.

Daisy insisted that potatoes had to be planted in the light of the moon, and tonight was the last night of moonlight in May. I walked back up the hill toward the back of the house. There, standing in the middle of the field, was a doe at the salt lick. I stopped, still as a cat. The setting sun burnished the doe's coat with golden light. She looked to be little more than a yearling, with delicate ankles and long, conical ears. I wanted to shoo her away, knowing that if Daisy saw her, she'd be tempted to get out the 30.06.

Suddenly the doe looked straight at me—maybe the wind had shifted and she'd caught my scent. She didn't bolt, though; just stood there, one hoof raised, as if she were ready to go but too curious not to stay a minute. I hardly breathed. Her head tilted first to one side and then the other as she inhaled the mélange of scents that came from me. I wondered if she could smell the venison we'd had for dinner. I edged toward the fence, wanting to see her better. And then I heard the surge of the chain, and Jiggs's *roo, roo, roo, roo* as he rushed out to confront her, all guard dog now. The doe went bounding to the edge of the field and sailed over the fence as if it were not there at all.

Daisy came running around the side of the house. "Was that a deer? Run get the rifle."

"I don't know what it was," I said.

I knew what came next. "Wish your mama was here. She could put a bullet into the heart of an elk a half mile away." She looked at me, standing there empty-handed. "I thought I told you to go get the turpentine. Sun 'bout to set and we gotta get these potatoes in 'fore the moon rise."

We worked in silence. Daisy plunged the shovel deep into the soil, slicing out wedges dark as devil's food cake. I followed behind as she slammed the dirt with the back of the shovel to loosen it, then reached down and pulled out the weeds. My hands were cold, and the work was tedious. At the end of the row, something writhed in my hand. In the failing light I couldn't really see it, but I could feel that it was cold, wet, thick, and alive. I screamed and jumped up, right into Daisy.

"What is wrong with you tonight?" She took me by the shoulders and backed me off her. I knew she didn't like me to hang on her. She'd had enough of that with her brothers and sisters.

I pointed at the ground. She looked down and laughed.

"Sweetheart, that ain't nothin but a night crawler." She picked it up. "Indians said these is some good eatin." It was a good six inches long when she stretched it out. She turned it over. "Run get that coffee can. We'll collect these for when we go fishin."

Once the worm was in the can, she sifted a handful of soil on top of it. The dirt lifted and shivered with the worm's panic. The skin on the back of my neck crawled.

"When Mr. Anderson was a buffalo soldier, he said them Indians sure did know how to do. Said he was huntin Apaches down in New Mexico one night, and it was getting dark exactly like it is now."

Oh good, I thought, another yarn about when she was a big rich rancher. We were going to be stuck out here forever, with the scary, slimy things.

"He was just fixin to turn back around when he heard something splash down by the river. He eased back in behind this big rock and lined up his rifle sight. Got it set in a niche in the rock just so. And then he heard it again. He swung the barrel over to the right and damn if he hadn't drawn a bead on an old Apache squaw with a little girl down by the river."

The sun was gone now and cold gripped the yard, but Daisy was sweating. She wiped her forehead with the arm of her checked flannel shirt.

"He said he cocked the hammer. Knew he was going to have to shoot 'fore the light went, and it was goin fast. But then he seen what she was doing. She was diggin worms, big as these. And then, so help me God, she rinched one in the river to get it clean and fed it to the little girl." Daisy paused. In the failing light, her pale-rimmed eyes seemed moonstruck. "He kept the gun cocked on them the longest. Finally, they moved on off and he put it down. He just decided he didn't have the heart to kill these two."

She looked down at the peck of potato pieces, as if suddenly aware of how dark it was getting. Behind her head, through the branches of the aspen on the hill, the silver moon pushed against the edge of the sky.

"Damn it. I'm 'bout to mess up this planting. Run get a pail of water so we'll have it to pour over these when we get them planted."

These two, I thought, as I struggled back from the well. How many others had there been?

~~~

Summer came on, the sky full of columbine now except when storms blew in out of the Arctic, with a fury that pounded the berries to mush. Minutes later, they'd vanish just as suddenly, leaving behind the smell of the rain, an aromatic balm on the land and all its creatures. Daisy loathed these storms, which sent her running to the back of the house, where I'd find her trembling between bed and wall, rocking back and forth while the lightning flashed and the rain pelted the sides of the house in sheets. I loved this dramatic weather, my euphoria compounded by the knowledge that I wouldn't have to water for days, but on those dark afternoons when thunder shook our little house, it made me uneasy that Daisy, with all her smarts, blubbered like a child.

The end of school meant no more library, and I spent most of the long days home alone. Daisy thought it was time I learned to work, and set me chores to do each day before she left: wood in, ashes out, water up the stairs, slops down. Gama and Gampa were down the road if I needed them. By seven thirty, I was on my way from Daisy's house, with Jiggs at my side. Gampa, an early riser, would have hoed the weeds from the entire strawberry bed by sunup. Gama would pour herself a last cup of coffee and make me a cup of sweet milk with cornbread crumbled in. I didn't much care for food in the morning, but if I finished, she sometimes let me have a watered-down half cup of sugared coffee too before I went to do my chores. After that, I was on my own, nothing to relieve the boredom of watching the steers fatten, the flies buzz, the pea pods fill—boredom so excruciating it made me build whole dramas around my little mudpies until even Jiggs would nod off.

Now it was time to move the geese from the shed to their summer quarters, a wood and wire pen down by the creek. Bess led the way. She hissed at Jiggs, her head low and menacing, when she approached the porch where he was chained. Then she and Amos signaled the all clear, swishing their tail feathers. They proceeded with pigeon-toed grandeur across the yard, their brood following obediently, sporting stiff new pinfeathers.

The goslings' schooling was immediately under way. Bess

worried and harried them until they cried out. The business of grooming gave them no rest. Their feathers required constant oiling to stay waterproof, from ducts along their tails. Down had to be fluffed for loft and insulation. Wings and tails had to be cleaned and preened for eventual flying. And constantly they had to be on the lookout for predators. If anyone other than Daisy or me approached, Bess stepped out in front of her brood and hissed. If the intruder continued, she raised her wings, lowered her head, and charged. In all the years I knew her, no skunk, weasel, or coyote ever harmed a single gosling. Each night they hunkered down beneath the outstretched wings of their parents, the gander facing north, the goose south.

When Daisy was home, all the geese's attention was focused on her. They watched as she unloaded the boxes of discarded produce she'd gathered at the market on her way home and waited when she disappeared inside the house, their eyes on the front steps. When she came back out, we began the task of sorting the produce. The bruised fruit Daisy would turn into wine, vinegar, baked goods, and conserves. The wilted lettuce and other leaf greens we threw to the hungry geese.

But when Daisy's Jeep departed down the road, the geese pressed their breasts against the slats in the fence that separated them from the creek, like prisoners listening to traffic outside the penitentiary walls. They didn't seem to know that there was no top on the pen, and that they could just fly out. Being penned in right next to fresh water yet having to drink and bathe from a tepid, filthy tub was a special kind of torture. Once the water in the tub was gone, they held their mouths open and panted, as though moisture might be gleaned from drawing in air.

All summer long I watched them. One especially hot morning I decided to let them out. It seemed there would be no harm in freeing them for just a little while. But years of exposure to water had swollen the wood and rusted the nails tight, and nothing moved when I tried to pry a plank loose. I headed for the toolshed and returned with the crowbar. I got one board off, but in the middle of the second, I stripped the head off the nail, lost my balance, and went sprawling backward into the creek. The shock of the icy water made me scream, which set the geese to honking. I was seriously reconsidering the wisdom of the Good Samaritan when I heard Daisy's Jeep shift into first at the bottom of the hill.

I ran in the house and tore off my wet clothes. Just as I got my Buster Browns tied, the Jeep pulled into the driveway, and I pounded down the stairs. I didn't notice my thumb was bleeding until I reached to unlock the gate. I must have laid it open when I slipped in the creek.

"I got through early, so I thought we could get some weeding done," Daisy said as she stepped down from the Jeep. She looked me over. "Why are you sweating so hard?" Then she noticed the crowbar on the ground in front of the goose pen. "What were you doin that needed that?"

I mumbled some feeble excuse about needing it for kindling and hurried to replace it. She let it go, although I could tell she was studying me as she picked over bruised strawberries to add to rhubarb pie. But she seemed to forget about it as we thinned onions and carrots until full dark. I took extra care with the hair roots, trying not to draw any more suspicion to myself.

I had trouble sleeping that night, with the ice-blue shafts of moonlight streaming in over the tops of the geraniums on the windowsill making the curtains fairly snap with lunar energy. The aspen leaves shivered in the wind, punctuated by the hooting of the owl who lived in the blue spruce down by the creek.

The well pump kept shutting off and switching on. Daisy snored thunderously in the twin bed next to mine—long gasping inhalations followed by snorts so intense they'd wake her. She'd lie silent then, momentarily stunned, and I'd hold my breath waiting for her to resume breathing. Then she'd snort again and her lungs would once more start sucking oxygen. There never seemed to be enough air in any room for both of us.

My thumb throbbed. I fantasized my jaw locked shut with tetanus, and Daisy's rage at my turning both stupid and mute. I was thirsty as dust, but it was always a risk to get up for a dipper of water. Daisy woken out of a dead sleep could be treacherous, especially given the 30.06 rifle she kept cleaned and loaded under her bed.

At breakfast Daisy admonished me, "You got no business makin play-pretties out of my expensive tools. Get the peas and beans weeded and watered. Fill up three buckets of coal. Bring the quart jars up from the cellar and get them washed up. Wash the peaches and get the bruises cut out. And stay out of that toolshed if you don't want to get slapped into the middle of next month."

I nodded contritely. I knew if I didn't water first thing, putting liquid on the plants in the middle of the day might boil them, and Daisy would know I'd shirked my chores. But before the Jeep had cleared the hill, I was back in the shed anyway. When I brought the tools around to the goose pen, I realized why I'd heard the well pump all night. In my haste to get back to the house the day before, I had left the hose running. The pen was a puddle of cold mud soup.

I pried loose the remaining boards blocking the geese from the creek. Bess, her bottom side coated with shit and spoiled lettuce, was the first one out. Limping and slipping down the bank, she lost her balance and had to run to keep from falling, ass feathers in the air, wings flapping, honking in alarm. She plunged into the shallow water, then righted herself and drank deeply. The bank confused the goslings, but once they reached the water they drank without hesitation, delicately dipping their beaks and then lifting them skyward to let the water roll down the full length of their throats. Sated, they began to bathe, slipping effortlessly into the water as if they'd come home.

I sat on the bank and watched them all morning, until the sound of Swinehart's wide-body pickup in the distance broke my reverie. The geese had left the water by now and had begun to roam into his pasture. To my relief he continued into town instead of heading up the road, but an awful realization had dawned on me. I had thought only about letting the geese out. It hadn't occurred to me that they would hardly thank me for their freedom by climbing docilely into the slimy pen. In a panic, I tried to herd them back across the creek. But as if one kind of freedom led to another, suddenly they lifted off as one body— a single thunderhead of gray, pounding feathers—and formed a perfect chevron, with Amos in the lead. I saw them land a quarter mile away, across the road from Mama's house, but I had no way of getting them back home.

What would I tell Daisy? The sun was halfway down toward Copper Ridge, which meant she would be home any minute. I tried to dream up some sort of acceptable story as I nailed the boards back on the empty pen, but nothing came. The safest bet was to deny everything. Our winter coal had just flown south, and that was that.

I hung the crowbar in the toolshed, at exactly the same angle I'd found it that morning, and hurled myself into my chores. I chopped kindling, washed the breakfast dishes, unpacked the canning jars. When the Jeep finally crested the hill a half hour later, I had the stove roaring with a coal-banked fire and the peaches sorted and ready to scald. By the time Daisy dragged herself up the stairs and into the house, carrying a peck of over-ripe plums, I was up to my elbows in sugared peaches. She covered the room in a glance.

"See you didn't get the beans weeded," she said, putting the plums on the buffet.

"No, ma'am," I mumbled. "I forgot."

"You didn't do the cabbage neither."

"I forgot that too," I said. I tried to sound normal even though my hands were shaking.

"Well, that's okay. Guess there was no water left after you did the goose pen. They sure did clean themselves up pretty today."

"Yep," I said. *What was she talking about?*

"We'll have to wait for the well to refill before we try to get any watering done. Anyway, it's too hot now. It will be nice to get out in the cool after sundown when we get finished putting up this fruit. Lord, I need to take me a drink of water."

Was this some ploy? I had to check.

"I'll be right back," I said. "I forgot to put up the posthole digger." Before Daisy could answer I was down the steps and on my way to the goose pen. There was the whole flock, lined up and meditating on the house in their accustomed places.

Then I got it. At the sound of the Jeep's return, they had all flown back into the pen, bathed clean as new spring. Hating Daisy as I did, I never counted on how completely they worshipped the woman who had first lifted them out of that railroad crate.

Hornets, yellow jackets, wasps in the eaves, restlessly indus-
trious. The creek, the trees, the bawling of a steer for no reason. Then too soon it was fall, the stunned leaves suddenly
golden against the white bark of the aspens and the still-cobalt
sky. The oaks rusted red, almost like holly. The birds hurtled
south. Each night the dark and cold came earlier, the smell of
woodsmoke on the wind wafting from the neighbors' ten minutes sooner. Less water flowed in the creek, and at night, a film
of ice formed along the edges. It was time to dig up the potatoes, their skins the color of apples, time to tie off the onions
and braid the garlic into gray ropes. The first snowbird appeared on the sill, the first rifle crack rippled across the valley.
Hunting season again.

Cold has a smell all its own. The basement wall was already
frozen, the crock of chokecherry wine heavy as stone. A deer
hung in the shed to cure, driving Jiggs mad with the scent of
fresh meat.

And so the seasons rode past, like a horse cantering downhill with the bit in its teeth, all my sulky resistance beside the
point. Until once more snow claimed the place, because it could.

A day the shade of old tinfoil. As I walked, the light bounced
off the snow in a hostile way, revealing nothing. Even though it
was nearly ten, the cold refused to let up. It could have been
seven in the morning or four in the afternoon. As I passed
through the chained gate, a weight descended on me. I heard
the grate of the iron plate being slid into place on top of the coal
stove. Gouts of inky smoke roiled up from the chimney, and the

same gray ash that coated the inside of the house silted down on the sludge-snowed roof.

Even before the door opened all the way, I could hear Daisy. "Uh huh," she was saying, as if we were in the middle of an argument. "Told you to take ten minutes and you been gone nearly an hour."

When she swung the door all the way open, I nearly gagged. Her stomach was saturated with dark blotted blood, and she seemed to be awash in a sea of reddened feathers. They drifted in waves about her, so that the floor appeared to float when she gestured.

An old black cowboy from Meeker had had good luck with his twelve gauge, and brought us a dozen mallards along with four snowshoe rabbits. Daisy was dressing the ducks, a phrase that never made any sense to me—they already had dresses. Now she waved a bloody finger at me. A feather was stuck to it. "Don't let out all the damn heat. Can't for the life of me understand why you want to stay out in the cold all the damn time." Directly behind her, over the table, was a picture of the sorry Jesus, savage thorn hat stabbing him in the forehead. Every time I saw it I wondered why he didn't just take it off.

I darted in past Daisy before she popped me. The stove was roaring so hot the chimney glowed coral, and the heat nearly knocked me down. Water bubbled in a tub on the stove—in fact, the entire house seemed to be boiling.

The ducks were piled on the claw-foot table, their beaks open as if they were in the midst of conversation. I wondered where their duckness had gone. Clay, the cowboy who had taken it from them, was a nasty root of a man but good at breaking horses. Every time he came around I had the feeling he was coming for me, even though he talked to Daisy cheery as could be. One day he'd dropped by when she was gone, and before I knew it he'd taken the hose and sprayed me. It had started out as play, but then he got a wild look in his eye and it turned scary. When I ran away, he'd stood staring after me for a very long time.

I stroked the head of the biggest male. The feathers gleamed dark turquoise, my favorite crayon color, oil-soft and smooth, and there was a white stripe around his throat. The lady duck wore heathery brown, as if she'd taken on the color of wheat or dried grass. Their legs lay apart and their feet had wilted like orange flowers. Daisy grabbed a handful of the female's breast feathers and ripped them out in one smooth jerk that made a sound like an old rotten sheet being shredded. It didn't seem right to look at the duck now, her skin a crosshatched grayish pink like the skin of a white woman. Daisy carefully tucked the wad of feathers into a canister by the table.

"What's that?" I asked.

"We fixin to make some fine pillows," she said. "I been saving a couple barrels of goose down, and with these, I think I can finish up a big quilt too." Down was the only thing that kept me warm at all—I was always cold. But I didn't like the idea that it had been ripped from the mallards, leaving them with nothing to wear.

Daisy picked up a couple of ducks in each hand and carried them to the stove. Holding them by their feet, she dipped them head-first in the tub of boiling water. She pursed her lips, blowing steam away from her face, then lifted the ducks and dipped them again.

"Why are you cooking them with the feathers on?" I asked.

"Ain't cookin them," she replied, focusing on the birds. "This is so we can pick the feathers and have them come out clean, without leaving in the quills." She held the ducks above the pot, shook off the extra water, and threw them in a dishpan. Working lightning fast, she yanked out the rest of the feathers, dipping her hands in a bucket of cold water from time to time. In less than a minute, the first duck was completely stripped— back, legs, wings.

I was both fascinated at her deftness and disturbed. Where had the duck itself gone? Was this offhanded deadness the same as my mother's? Did deadness happen so casually to all things? If I let the word *Mama* come into my mind, I would start to cry. It always happened. *Mama.*

The steaming feathers stank. The windows of the small house started to sweat, and I felt the hair on the back of my neck stand up as Daisy dipped the ducks in the boiling water, this time feet first. Then she began to peel off the skin.

"Hand me that cleaver."

I didn't move. How could the ducks paddle around heaven without feet?

"Dammit, heifer! You hear me talking to you."

The square steel gleamed dully next to the dark, bloody carcasses. Gingerly avoiding the stained handle, I picked it up and carried it to Daisy. She placed the duck on a wooden slab and brought the blade down hard through the knee joint. I jumped. The foot separated cleanly. She tossed it into a pan that was nearly full of the Halloween-orange feet, with their tiny little toenails.

"What you so jumpy for? It's time you get shed of this skittish streak once and for all. We don't throw away nothing here. And duck feet soup is delicious."

I didn't know what to say. I couldn't think of anything that bridged between Daisy's world and mine.

She went on. "Yes, Lordy. We'll get three, four meals out of these. Pan-roast up some sage and make up a big mess of duck sausage. Fry the livers for breakfast. Roast the rest of them, two at a time. I can eat a whole one by myself."

I knew better than to say what I was thinking. *I won't eat any of them.* But she must have seen the expression on my face.

"You know how you so crazy 'bout chicken drumsticks? Where exactly do you think they come from? You must got that from them silly town kids think meat comes from a store."

I looked up at her. I wished I lived in town because no matter how many animals I saw killed and dead, I just couldn't get right with it. Maybe it was because I felt so keenly the fulcrum of death that our own lives seemed to teeter on. I hated to further the hurt that seemed to be all around and in me.

Right on cue, my eyes began to water.

"Rita Ann. Straighten up. I ain't got time for that pissin and moanin now."

I felt ashamed, ashamed and adrift. A bum lamb, weak from the start. I couldn't stop crying, though.

Daisy grabbed me by the hand, dragged me back to the table, picked me up, and sat me in a chair. Then she brought the pan of naked ducks, spread out some butcher paper, and tossed them onto it.

"This here's dinner. And we're lucky to have it." She slid a long knife along the sharpening stone, wiped it with a rag, then stuck it into the duck with a practiced stroke, just above its tail. The guts spilled out cleanly, the entire mass in one handful.

I stood up so I could see better. The intestines were surprisingly bloodless, but they smelled gamy. "What's that?" I asked, as she cut something free of the coil of bluish tubes.

"This is his liver. And here's the heart and the gizzard."

"Does everybody have guts?"

"Everybody," Daisy said, concentrating. "But only birds have gizzards." She sliced open something that looked like a clam. "See here, this sand?" She scraped it onto the paper. "He uses sand to grind up his grain." She picked up the liver. "You gotta be careful with the bile sack. If you break it, it'll taint the flesh." She placed the liver carefully in a white bowl along with the other livers and hearts.

She rinsed the gutted duck in a pail of water. "In the South, they wouldn't let none of us have no meat when we was share-croppin. If we hadn't been able to slip off and hunt once a week, we wouldn't have had any at all."

"Who would want it? These stink." The words were out of my mouth before I could think better of it.

Just as rapidly, Daisy slapped my face, hard. "You simple? Where in the hell did you come from anyhow? I know you ain't none of Mae's child. Your mama was a go-getter, wasn't nothing mealymouthed about her. You must got this streak from your dad's side of the family. Well that don't make no never-mind. I aim to break you of it."

She'd used that expression before, and I always wondered if it meant she was going to crack me in two, like a plate. Or cleave me, like the ducks' feet. Fear settled like ice in my stomach.

She pushed the rinsed duck toward me. "Now this is your job. You got little fingers, so what you must do is reach up in there and feel around for buckshot. Else we be in a heap of hurtin if we bite down on a pellet of lead."

Daisy loomed. The smell overpowered me. The heat pressed down. I continued to sit on my hands.

"You do not want me to get down that strap. I am not play-ing with you." Her eyes were hard now.

I took one hand and slid it inside the body of the duck. It was hot and wet, and the bones were sharp. It felt like reaching inside a little house. I passed my fingers along the ribs. I won-dered if when spring came and we were back at the lake, the new ducks would know what I had done. Still, I was curious. I pressed my hand against the breastbone and there, smooth as a little rock, was a pellet of lead. I hooked it under my fingernail and brought it out.

I lifted it up so Daisy could see it.

"That's the very thing. Now why did we have to go through all that confusion and carryin on to get you to this point?"

She smiled. In spite of myself, I smiled back.

4 ~ Geronimo

*Robert Ball Anderson with his new
bride, 1922*

~~~~~ **A little wooden building** badly in need of
paint, the library where I waited while Daisy cleaned the church
was my first place of worship. I'd barely notice the hours fly by
until the little bell on the door would tinkle and I'd hear Daisy
calling, "Where's Reeter?"

At first I was only breathless to get to the comics, but as my
reading took flight, the grandmotherly librarian directed me to
books that ranged far beyond the fairy tales I had at home. I was

obsessed with horses, and embarrassed that when my ranching classmates talked of fetlocks, flanks, and withers, I wasn't exactly sure what parts of the animal they were referring to. I wasn't certain what a bay looked like, and I was far too proud to ask. But the librarian introduced me to the encyclopedia, which seemed to know everything there was to know about horses. There was a diagram of a horse, with all the parts labeled, and a picture of a bay, which I realized was the common type I saw across the road every day.

School was no church to me, but it was another refuge from Daisy and her mood storms. At reading time I could cocoon myself within a story so completely that "I" ceased to exist. Never mind that no one in the stories looked like me, a horse-loving Western black girl with neither horse nor ranch. My entire consciousness was subsumed under the kindly gaze of the nice policeman, the competent dad, the understanding mom in the pages I turned.

The rest of the time, I wondered at the odd subjects they taught. Why would anyone diagram a sentence? I couldn't understand this business of breaking things down into their particulates. Outside in the garden, we were always finding arrowheads, but Kawlija, the wooden Indian from the country song, was the only Native American I heard much about. In my history books, no mention of lynching or blood in the soil; in social studies, no discussion of how one group forced another to do its laundry, and to keep on doing it for so long that the first group forgot they had ever had dirty underwear. I wondered how one broke free of the laundry-doers and joined the laundry-shirkers. That interested me far more than subjects, verbs, and predicates.

But I fell in love with science, because of my first male teacher, Mr. Mason. He had a perpetual tan that made him look well-to-do, and an equally perpetual sense of humor and easy, open manner. I loved to be near him. Even his flaws endeared him to me—the shiny spot on his head where his hair was beginning to thin, his habit of coming to school with shaving nicks plastered over with tissue. At night, I did my homework for him religiously, and when I slept, I dreamed of him.

Science fed my insatiable hunger to know about real, measurable things. Once Mr. Mason had us do an experiment where we linked hands while electric current flowed from one salty, sweaty palm to the next. Another time he showed us a film about the demise of the poor dinosaurs, whose God had failed them so capriciously. I was sad for days, thinking as I filled the coal bucket how cruel it was that the forests that had been their food were now warming our feet. But even when what I was learning disturbed me, I trusted it much more than the conundrum of redemption and blood the priest intoned at church. Drink Jesus' blood? Why would anybody want to do that? And why would Daisy worship the very angels who had made off with my mother?

She was unshakable. "They got wings, and they live in heaven with Jesus, and they carry out the Lord's orders."

"Wings? Like bugs?"

She gave me that cut-it-out look.

"Not like bugs. Wings like a bird. Like the Holy Ghost."

But the room where they kept the Holy Ghost didn't seem to know about Mr. Mason's classroom, where we looked at frogs' guts. It kept me up nights, trying to reconcile the two. My mind seemed trapped in a hallway between these worlds that did not seem to intersect anywhere.

One day one of the students brought Mr. Mason a gift, a shot glass depicting Indians who lived high up in the walls at Mesa Verde, where the student's family had gone on vacation. Mr. Mason gave it a place of honor next to his pencil cup.

I asked Daisy whether I could bring Mr. Mason a gift too.

"Why certainly," she said, ever eager to curry favor with my teachers. "How about a couple of fresh pullets?"

Something about the suggestion seemed a little weird, but I wasn't sure why. Then I had an idea. "What about a loaf of oatmeal raisin bread?" Daisy made great oatmeal raisin bread.

That weekend she slaughtered a couple of young chickens by her usual method: She held them upside down by the feet, secured the neck under her shoe, and pulled until the head came off. Then she gutted and plucked them and stuck them in the Deepfreeze. Sunday she made the bread, feeding the yeast sugar and warm milk and setting it at the back of the stove so it could grow, then cutting the cinnamon and sugar mixture with walnuts so the bread would be the perfect mixture of sweet and chewy.

Next day I got on the school bus all aquiver, carrying my presents—the bread, which smelled so good my mouth watered, the two worrisome chickens, a jar of apricot jam Daisy had thrown in and a note she'd written to Mr. Mason. By the time I got to school, though, I was starting to feel weird, picturing these messy, excessive gifts next to that little shot glass. All through morning classes, I waited for a moment when no one was crowded around Mr. Mason's desk, but he was the most popular teacher and that moment never arrived. Meanwhile, I was growing anxious about the thawing chickens. Morning recess came and went, lunch as well. I spent afternoon recess in the bathroom, and just before the bell rang, I lobbed the sack onto Mr. Mason's desk without the note Daisy had written and ran back to the bathroom. I waited until I heard the rest of the class coming up from the playground, then headed straight for my desk, pulled the lid up, and busied myself with my pencils.

"*Eeeek*, it's a raw chicken!" shouted one of the kids.

I wanted to sink through a hole in the earth. I stole a glance at Mr. Mason, who had been writing something on the blackboard and hadn't noticed the plastic bag on his desk, or the red-tinged liquid leaking from it.

"Whose is this?" he asked, frowning. He clearly didn't realize it was a present.

The other kids had gathered round his desk, poking at the bag. "What is it? What is it?" I forced myself to join them rather than hiding in the back.

"It's *two* raw chickens," said Ronnie, lifting one corner of the bag with his pinkie. I forced myself to resist the urge to bolt out the door.

"And bread!"

"And jam." This from Vicky, who lived down the lane.

I could feel my face flaming. *Please, don't let them figure out it was me,* I prayed.

Mr. Mason looked around the room. "Who did this?" When his gaze fell on me, I did my best to look as wide-eyed as everybody else. When no one answered, he scooped the bloody sack into the trash and then, holding the wastebasket out in front of him so he didn't stain his immaculate shirt, hustled it out of the room.

"Yuck," I said loudly.

~~~

Daisy wanted to "make something" of me, and I never knew quite what bizarre form her ambition would take next. Once I got off the school bus to hear singing coming from the house, in a language that wasn't English. I went inside to find a young man with frantic acne and disciplined hair sitting at the table, wolfing down a loaf of bread and a pot of jam.

"Tony's going to teach you German," said Daisy, nodding at him. "Ja, I can teach you Cher-man," he said, his lips glistening with butter. And so, for an entire year, he did, setting himself loose upon the lyrics to songs from his homeland—"Die Forelle" and "Der Handschuh" and "Der Erlkönig"—with a homesickness beyond anything I had ever witnessed.

Another time I arrived home to find a piano in the house, tall as a coffin against the wall. The lid opened like a mouth, revealing eighty-eight black and white keys. "Mrs. Gilbert is going to give you lessons," Daisy said. "And then you're going to play for the church. I'm telling you now, no honky-tonk."

Mrs. Gilbert's parlor smelled of old roses. I plucked at the edges of my notebook, marveling at the aging stacks of music piled along the floor, the dark banister curling down the chairs, the formal wallpaper shrinking a bit at the seams. And then Mrs. Gilbert appeared, an elderly bespectacled pigeon smelling of rose water. I had to check the urge to throw myself into her lap and say, "Please be my mother and let me move in here."

"Hello, dear," she said, wheezing softly. I followed her down the hallway into a room where a warm yellow light cast a glow on the keyboard of a glossy black upright. She motioned me to sit on the bench and seated herself to my right, holding a thin metal rod she called a "baton." My legs didn't reach the floor. *Every Good Boy Does Fine* marked the lines of the treble clef, *F-A-C-E* the spaces. What was I doing here, on another mission that seemed urgent to Daisy, pointless to me? She opened a book to "Home, Home on the Range." Why, I wondered, did they make a song about that?

And yet week by week, the music began to lurch out in miraculous little fits under my hands, astonishing me. I leaned close to the notes sprayed across the page as Mrs. Gilbert beat the edge of the piano with the baton. I hadn't expected that I would be able to make the thing work but here was the music, note by note, willing to play along.

Hanon, exercises to strengthen my fingers. Octaves and scales to make them agile. Mrs. Gilbert took my hands in hers. Tuck the thumb under without lifting the wrist. Cup the hand like you're holding a ball. Hit the key with the tip of the finger. "Sit up straight. Why are you leaning in so close to the piano?"

At home in the evenings, Daisy sat by me with a piece of freshly split firewood, bleeding sap. "All right now, no mistakes." She didn't understand the concept of practice. Do it right from the start. Piano as one more weapon in the dreary war against racism. And so music, which had seemed at first to invite me in, was reduced week by week to yet another hurdle that had to be mastered for Daisy, that could have no meaning for me.

Then one night when Daisy had lingered behind to pay Mrs. Gilbert, sending me on out to the Jeep, she came out so furious she wouldn't even look at me.

"Daisy, what's wrong?" I asked.

She only started the engine and popped the Jeep into first. At the market she barked, "Stay in the car." Same routine at the post office. I wondered what it could be. I'd had an unremarkable lesson. Mrs. Gilbert had begun to introduce me to simple compositions by Beethoven and Schumann. She'd encouraged me to think about what I would play for the recital. It wasn't until we reached the base of the hill to Strawberry Park that Daisy expoded on me.

"So why didn't you tell me you couldn't see?"

That familiar leaden panic. What did she mean?

"Mrs. Gilbert told me you can't see to read the music. Make me look like I don't take care of you."

"I didn't tell her you did anything."

"You callin me a liar?"

"No," I said, hating her fully.

And that was how we learned that I was legally blind in one eye and needed glasses.

~~ ~

It was becoming increasingly clear to me that no matter how fast I played "Für Elise," I was going to remain the issue of a union Daisy regarded as a tragedy and a disaster. Living with her was like riding an inner tube down icy rapids. Skimming along above treacherous boulders in a setting as bucolic as a Norman Rockwell calendar, I had to concentrate all the time just to stay afloat. I couldn't afford ever to relax; I had to try to anticipate what was just around the bend.

I can't feel this now, I'll feel it later, I told myself when the harshness of her spirit began to wear me down. It seemed to work. I could forget anything, or so it seemed. The problem was, I often couldn't remember things I needed to, like seven times seven, or where I had put my coat, my schoolbooks, my pencils. What I did right hardly mattered—only my errors registered with Daisy. Desperately I tried to invent a Rita who could skim above it all. I developed an accent, pronounced "tomato" *tomahto* and "privacy" like I meant to say *privy*. My classmates laughed at me. Daisy ignored it altogether.

When school ended, the library ended with it. I was thrown back upon Daisy, and what little escape I had at home. My cheap cardboard editions of fairy tales, with their pretty princes and princesses, had begun to fray in more ways than one. Like the princess with the pea under her mattress, I was beginning to be bothered by the similarity of all those Snow Whites, left waiting "in a coma," as Daisy put it—in a state of suspension with which I was far too familiar. Still, eventually the prince came for Sleeping Beauty, on his long-maned pony. He came again for Cinderella, with her tiny, fragile shoe. Why had no one come for me?

Was it because I wasn't white enough? There were no dark princesses or fairy godmothers. Dark was malice, mischief, evil, and death. It hadn't escaped my notice that each night before she went to bed, Daisy applied her bleaching cream, even though every morning she woke up looking exactly the same. She pressed her hair and mine with a hot steel comb until it wilted flat as straw, although the slightest bit of moisture in the air made the curl spring right back. I'd never heard the word "nigger" outside of home, but inside I heard it every day.

"Nigger, what's the matter with you. Got your mouth all puffed up."

"A nigger's a nigger and a mule's a mule, and you can't make nothing out of neither one of them."

"Don't never get round them city niggers with that light complexion. They'll cut your throat in a minute."

It was an acid that never ceased corroding.

Outside in the pasture, the unconcerned steers continued to grind their cuds. Across the road, horses in five different colors munched the splendid grass. Daisy kept rattling on, spinning yarns of murder and mutilation while we hulled peas into a dishpan, the air perfumed by night-blooming jasmine.

"Mr. Anderson said they cut off their titties and tanned 'em and made tobacco pouches."

"Who?"

Daisy looked at me as if we'd never been introduced, blinked, then huffed, "I was talking to myself. Mind your own business."

But the times I tuned her out, I'd get in trouble too: "Don't you hear me talking to you?"

I was collecting strands, trying to sort out what to ignore, what might be true. Or more important, what might be true for me. I went to see the movie version of *Snow White*. When the evil stepmother asked the mirror, "Who's the fairest of them all?" I flinched. I was fairer than Daisy. Was that why she didn't love me? Would people fairer than me reject me as well? Though I was darker than anyone else in the third grade, it had never seemed to matter in the classroom or on the playground. When I played jacks with my friends, the best one won. Period.

Except: There was one other dark kid in my school, a hand-some Italian boy in the sixth grade. I had never so much as spoken to him, but I developed a crush on him that I unwisely confided in my friends. One day after recess he rushed up to me. "Don't you ever again tell anybody you like me," he spat. "I don't want anybody to think I'd go with a nigger."

It was the first time I'd heard that word spoken outside home. I didn't tell Daisy what had happened. I didn't want to confirm her sense that "they" considered us to be dirt. Nor did I want to reveal that I had failed somehow to accomplish the mission she had set out for me—that of being so pure, bright, and fair that no one would ever apply the old labels to me. But I became wary. Who had told him I had a crush on him? Who else thought of me as a "nigger"? He'd sounded exactly like Daisy when he said it, his voice full of loathing and his face con-torted with hate. It was as if some depth charge planted long ago had detonated.

Was this thing that afflicted Daisy something that had at-tached itself to her shoe when she tried to escape the South? Was she crazy, or did she know something I just hadn't figured out yet—something everyone else, black and white, knew full well but never discussed? Was all this work to make good grades and learn the piano and German going to turn out to be a waste in the bigger world merely because of my hide? Were people really that simple? Feeling traitorous, I hoped that Daisy's darkness would not attach itself to me. Maybe on my own I could slip into the castle.

I flipped through the two books Daisy kept in the buffet. Robert Ball Anderson's memoir was a small paper-covered volume. "I am a full-blooded African Negro," he declared—or Daisy declared for him. "There is not one drop of white blood in my veins." Supposedly Mr. Anderson had nearly died as a boy, of a beating administered to him by his alcoholic mistress. Two years later, when she got down the whip a second time, he yanked it away and applied it to her. Or so the story went. I wasn't sure I believed it, or any of the stories Daisy told about the past. Still, how brave he sounded—and how resourceful. When he'd decided to leave the plantation, he'd just up and gone. Out west in his days as an Indian hunter, he'd had to share a blanket with a lice-infested soldier on a patch of bare ground. Daisy said they'd covered themselves from head to foot in a coat of wet mud to suffocate the lice.

But it was the other book that really captured my imagination—a biography of Geronimo, its musty-smelling, yellowed pages bound in red cloth. What it was doing in Daisy's buffet, I had no idea. Mostly it was about Geronimo's adventures fighting the cavalry. But what intrigued me was exactly how he avoided them. He knew to step on soft grass, in tracks that had already been laid down. Like Mr. Anderson, he'd found a novel use for mud. One night when the cavalry was hunting for him, he covered himself with it and let it dry. Come morning, he had merged with the landscape, looking like just another pile of dirt while the cavalry hunted all around him, the horses' hooves actually brushing his hand once. How strong he must have been not to move for so long! When the unit finally gave up, he stood up, shook himself, and loped off.

If Robert Ball Anderson and Geronimo could light out for the woods, why couldn't I? I'd seen a tepee up at Elk Park once. I could camp out there with Jiggs. We would drink from the irrigation ditches. Maybe we'd have a problem with bears or bobcats, but that would be Jiggs's department. He didn't even like it when Daisy raised her voice to me. "You done ruined that dog, turned him into a play-pretty," she complained once when she came home and discovered I had tied a bandanna on his head.

The idea of this adventure electrified me. Daisy was off cleaning, my grandparents were down the road. I started scrambling to put together provisions for our journey, determined to outrun the voice in my head that might try to talk me out of going. I got a sock from the drawer and stuffed it with cold pancakes and trout left over from breakfast. I considered taking Gampa's .22 rifle, but it was too heavy to lug around. I found the key to Jiggs's padlock under a geranium pot. When I came out the door, Jiggs went crazy barking, ears up on point. He could tell from my demeanor something was up.

We took off west, Jiggs running ahead then coming back to urge me to go faster. Past the poultry sheds where the geese, ducks, and chickens scootched down in the dirt, taking their dust baths. Past the outhouse and the old Ford Daisy had driven from Nebraska. At the end of the property I allowed myself to glance back, just the once. The houses behind us looked so small, almost dainty. For a second it crossed my mind to turn back, but Jiggs plunged ahead, down the embankment toward the Rudolph ranch, and I followed.

The hay came nearly to my chest, big blond heads full of grain. I began to sniffle and wheeze, but Jiggs looked back, urging me on. We plowed deeper into the field, disturbing lunching grasshoppers. Jiggs leaped up, snapping at them, his fluffy champagne tail collecting burrs as he went. My eyes started twitching. I rubbed them a little and then an unbearable itching came over me as the twitching began to tingle. By the time we got across the pasture, my eyelids had nearly swollen shut and my whole face had begun to seize. Had something bitten me?

Thankfully, at the bottom of the hill, a creek flowed by a corral. Jiggs plunged in and I dunked my head under water. Jiggs lay down on the muddy bank to cool himself. It occurred to me that I might dab some mud on my face, Geronimo-style—now that I was in the adventure—to calm my eyes and nose. Immediately it soothed me.

Jiggs's head jerked up on point, dark nostrils quivering. I leaped for him, but he dodged me. Once he was loose, Jiggs would not be caught until he was good and ready. "You can't do this now," I told him. He glanced at me, leaned down to lap some more water, then studied the corral behind us worrisomely. A rancher wouldn't hesitate to shoot him. But he loped off purposefully, nose high.

"Jiggs, you get back here right now," I said. He didn't even turn his head, just kept trotting toward the corral, where no cows were visible. Then I saw what he was after. He slipped under the fence and made for a pile of fresh green manure. Cow manure had a peculiar appeal for him that I didn't get at all. He rolled on his back, muddy paws in the air, groaning with pleasure.

"Jiggs, come on." He was going to make me beg. I shook the food sock at him. He cocked one eye at me, clearly delighted. Then, in his own good time, he stood up, stretched like a pussycat, and ambled back to me, stinking like a septic tank. If anyone came hunting for us, they would smell us a mile away. When I tried to pull him toward the creek for a bath, though, he spun out of my grasp, tail in the air, ready for a game of chase.

"Let's go, Jiggs, before you get us in trouble," I said.

We made it past the Rudolph house without seeing their livestock, their dog, or any of their eight kids. At the road, we headed north toward Elk Park. Here the horseflies and deerflies, drawn by Jiggs's aroma, descended without mercy. And this road that had seemed so flat from the seat of the Jeep turned out to have a steady uphill grade that was tiring, especially with the sun high overhead. I began to regret I hadn't brought a canteen. I picked up a stone to suck, reassuring myself that I knew a trick or two about survival. But then we rounded a bend in the road and the forest rose up, tall, dense, and imposing. The plants seemed utterly unfamiliar, and they certainly didn't look like dinner. I wished I'd paid more attention when Daisy harvested wild greens and berries. When Jiggs charged a little garter snake meandering across the road, I squealed and turned tail, rabbiting up the hill and into the woods. I didn't stop until I tripped over a root and fell, cutting my knee. I rolled up my pant leg, trying not to cry. Would it be easier for a bear to smell me now that I was bleeding?

Jiggs came running after me and licked my face, then plopped down and proceeded to chew a foxtail out of his muddy paw. I calmed down some. It seemed a good time to have a snack. The trout tasted delicious but they were surprisingly small. Jiggsie inhaled his, bones and all. I curled up, cradled my head in the basket of a tree root, and listened to the soothing wind high above. Next thing I knew, a pine cone bounced off my head, waking me. Sitting up, I could no longer see the sun, but I could hear horses' hooves. I grabbed Jiggs by the scruff of the neck and wrapped my legs around him.

When the riders came into view, I saw that they were just a couple of tourists on day-rental nags. I waited until they'd plodded back down the road, then found a walking stick and set off at a fast clip. But beyond every turn was still another turn, and then another after that. Jiggs plodded along with his tongue hanging out, panting, no longer galloping ahead or stopping to explore intriguing scents. Occasionally, he would glance up at me quizzically, as if to say, "Are you sure we should be doing this?" A vicious set of blisters had sprouted where my heels rubbed against the wet leather of my shoes. I was starting to feel just a little bit lost.

Then, behind us, I heard men's voices and horses blowing dust out of their nostrils. How had they come upon us without Jiggs letting me know? Irritated, I grabbed him and pulled him down beside me in the dry ditch, covering us both with the long thick grass. I kept one hand firmly on Jiggs's muzzle to keep him from barking. The men came closer.

"Well, she had to come this way. The broken-down hay from that house across that field led straight to this road." They had to be talking about us! Panic flooded through me. How could I have been so stupid as to leave a clear trail? Geronimo had said to walk in a path that had already been made.

I peeked up through the grass. Two men with white hats and stars on their chests, riding horses loaded with a good forty pounds of tack, and rifles. The sheriff's posse! They were systematically searching the far side of the road for footprints. A huge horsefly zoomed by and Jiggs snapped at it. One of the horses caught the movement and looked in our direction, and I

realized what good luck Jiggs's earlier misbehavior had been—
the manure masked his dog smell.

Both men reined their horses up sharp. "You hear that?"
said one. I held my breath and pressed myself down into the dirt
so hard the rocks gouged my spine. My scraped knee throbbed,
and Jiggs's panting seemed deafening. Nothing happened for a
space that felt like an eternity. Then they reined the horses toward us. The face of a big buckskin loomed above us, reaching
down to crop grass a foot from our heads.

"Must have been a squirrel."

"Yeah, I mean, she's just a little kid. Surely couldn't have
made it this far. That's nearly two miles." Two miles! We had
gone only two miles? Above us, the sky seemed to be closing in.
I fought a sneezing fit. Lying in the grass was bringing back the
hay fever, and my throat was dry as cloth.

"Well, maybe we ought to backtrack." I heard the sound of
a canteen lid being unscrewed, and then the sound of gulping.
"You want some?"

"Nah. You're probably right. I sure hope she didn't fall
somewhere."

The hooves receded, and we lay motionless until the birds
resumed their song. I looked around me. The forest had revealed its indifference now, and the virtues of this adventure
were diminishing. But we couldn't turn back now. Daisy would
kill us both.

I trudged on, making an effort to keep to the gravel, so we'd
leave no further trace. The turnoff couldn't be that far. At the
beaver pond, there would be sarvis berries and thimbleberries
and a wide field of dandelions, which I knew were very good
with bacon grease. If we could just get to the Whiteman School,
maybe we could find a hose to drink out of. But the light came

lower through the trees already, and my clothes were damp and muddy. One of the blisters on my right heel had popped, and every step I took stung. By the time I reached a footpath that seemed to lead in the direction of Elk Park, I was tired and discouraged, not like a wily Apache at all.

The path was steep, and now mosquitoes took over where the flies left off. Jiggs kept sitting down to chew at his right front paw. "Come *on*," I said meanly, and immediately felt sorry. This mess wasn't his idea to begin with. I opened the sock, fed Jiggs half of the last pancake, and finished the rest myself. Then I let him chew the trout grease out of the cloth.

As we made it over the rise, a black border collie with a white stripe down her forehead appeared at the edge of a grove and barked once. Then I heard a sheep's bell. The flock couldn't be too far off. Jiggs's ears pricked up, and his body trembled as the other dog inched closer, her elbows so low they almost dragged the ground. I stopped. Would she hurt us? Where was her master?

Jiggs threw his plume tail in the air and stuck his nose out, trying to inhale her scent as she circled us both. When she looked at me, I made a point of studying the mountain, to make it perfectly clear I was not a kidnapper of sheep. She looked at Jiggs sideways, as if the cow smell put her off—she was running sheep in cow country, after all. Then they danced around each other and she took off. Jiggs raced after her.

"Jiggs Williams, you get back here right now," I yelled to his disappearing hind end. I fought back tears. "Jiggsie," I called, but he didn't even turn his head. Disconsolately, I plodded after him.

Off to the side of a stand of pines stood a wooden sheep wagon the general shape of an old Conestoga. I knew where

I was now. This was the warm-weather home of one of the old sheepherders who came through the passes every spring, bringing their flocks to graze in the high pastures. I had heard they were Basques from the Old Country, but folks in town claimed they were just Mexicans trying to act high-toned. The ranchers called the sheep "Rocky Mountain maggots," a name I couldn't say—it was bound up with the writhing mass, bone-white and blind and busy, I had stumbled upon one day in a barrel where Daisy had thrown the guts and feathers of a few dozen chickens. But my disgust colored my attitude toward the herders and their herds, who the ranchers said ruined the range, eating the grass down past the root so that it wouldn't grow back.

I knew this herder, though. Daisy visited him to barter peach jam and venison stew for the sourdough starter she used for pancakes and bread. She kept it in a crock in the corner of the kitchen, where the smell of it quietly fermenting under its tea towel gagged me when I passed. It died surprisingly often, sometimes even of natural causes, but Daisy always insisted on replacing it.

The herder stepped from behind the trailer. He was as small as a girl, with a washrag of a hat and grubby fingers. His face was tanned as a saddle, his soil-brown eyes steady but kind. "Your dog?" he asked, pointing to Jiggs. He smiled, revealing teeth as blunt as his sheep's.

"You got any water?" I asked, pantomiming drinking from a glass.

He smiled wide and stepped over to a bucket on the step by the wagon. When he brought me a full dipper, I had an impulse to climb up into his lap and stroke his soft beard. Then through a stand of tall pines behind him, I saw his horse, a gelding so beautiful he could have been sired by the Black Stallion. The herder saw where I was looking and put his fingers to his lips. He whistled softly, and the horse's haltered head came up. As he came toward us, I saw that his back legs were hobbled. I put my hand out and he reached to sniff it, his muzzle as soft as Gama's apron.

The herder took off the hobbles and formed a stirrup with his hands. How did he know that when I was on a horse, I respected myself? I had dignity. I was part of the spirit of the West, not some castoff of Daisy's, shivering and cowering and mute. I stepped up, and he boosted me onto the horse's bare back. The gelding was surprisingly wide, and I had to squeeze tight to settle on. The herder began to lead him back the way we had come, Jiggs and his new friend trotting alongside.

At first I preferred not to think he was leading me anywhere special. But when we reached the main road, I said, "I don't think I want to go this way." The herder turned around and nodded as if he understood exactly what I was saying, but he didn't break stride. I didn't know what to do. We plodded along silently in the waning light, and an unexpected calm descended on me. Then I heard the sound of hooves coming hard.

Into view came a plain-faced woman whose family lived in Mama's house now. She was riding a big bay. "Hi, Rita," she said, with uncharacteristic neighborliness, pulling up close. "What happened to your face, hon? Are you okay?" I imagined how I looked, my face bloated and my braids undone on one side, so that my hair stuck up like fright itself. I was bruised and bitten, my jeans were torn and bloody, and I was so hungry and tired I could barely stay upright. I knew a beating awaited me, but I was secretly relieved to be found.

"Rita, if you'll come back, I'll let you ride my horse," she said, as though I had a choice. The bay was an old friend for whom I had pulled fresh grass every summer since he'd been foaled. "Okay," I said, embarrassed. I knew they had me. The herder helped me down, ruffled my hair, and turned to head back up the road. We set off in the other direction, and Jiggs, besotted with the border collie, raced back and forth between us as the distance widened, until he finally decided his place was with me.

I rode behind her, silent, and after one or two attempts to get me to talk, she gave up. I was thinking about the time I'd broken into her house when her family was away. On the doorstep I'd hesitated, my heart pounding. Something told me that if I went in, I'd obliterate the memory of Mama standing there, coming back from feeding the chickens, the sunlight glinting off her glasses. But the new reality had beckoned. I'd opened the door, only to discover a jumble of coats and swimsuits on the floor and an unfamiliar, perfumy smell I couldn't escape fast enough.

I was dismayed to discover how short a distance we had come. In less than an hour, we had reached the stretch of level road that ran to my house, where at least thirty horses and riders were gathered. I recognized the sheriff and his deputies among them. Were they going to take me to jail?

And there, at the center of it all, was Daisy.

"There she is," somebody called, spotting us. And amid congratulations and cheers, I climbed down into Daisy's arms. For once she hugged me, and I hugged her back.

~~~

When salvation came at last that summer, it arrived not on horseback but on drumbeats. Part of Daisy's summer job was cleaning at Perry Mansfield, the summer arts camp Charlotte Perry and Portia Mansfield had created in the mountains decades earlier, and sometimes she took me along.

I loved Perry Mansfield, or PM, as we called it. No matter where you went, the smell of pine sap followed. Portia and Charlotte—or Kingo, as she was called—had been drawn to Colorado for the cool summer air in those pre-air-conditioned days, and a spirit of rustic simplicity endured. The cabins and studios were composed of the barest scraps of lumber, with only tin and screen between people and the surrounding spruce and aspen. All the cabins were named—Sage Brush, Aspen, Willow. So were the cars. There were horses too—equestrianship was part of the curriculum. The sound of piano was always on the wind, tinny and out of tune, always accompanied in turn by a teacher yelling, as if running alongside the music, "And ONE two three, TWO two three." Then the brush of toe shoes across the wooded floor, and thudding lunges. I had no idea that the people wandering in and out of the studios in their dirty jeans included the likes of Merce Cunningham and Martha Graham and Dustin Hoffman, but I knew they were up to something magical. Art class in Steamboat was Elmer's glue, construction paper, and crayons. At Perry Mansfield I found out that art included painting something that did not look exactly like a horse or a bird—it could be just a splat on a canvas. I discovered there was such a thing as theater, and dance—and that those were art too.

I'd never met anyone like Portia, who dyed her hair the color of rust, or Kingo, a coal baron's daughter who clad her tall, slim frame in fawn-colored slacks and pale, tasteful blouses. The first time I saw Kingo, Daisy and I were cleaning the theater, trying to be as quiet as possible because a rehearsal was in progress. A boy and a girl who looked to be no more than fourteen faced each other on stage, speaking stiffly to each other in loud, unsupported voices.

Suddenly Kingo bounded out of her chair and up the stairs

to the stage, her frizzy, mushroom-colored hair bouncing. Andy, her standard poodle, with hair the same color, bounded up too.

"Stop, people," she said, waving her hands in the air. She sucked air through her teeth, studying her students dolefully. "You two are standing there waiting to say your lines, not listening at all. *The Cherry Orchard* is about reactions. Remember, the serf now owns the entire property he used to serve on, and it is his pleasure to invite his former masters to continue the party." She turned to the rest of the campers seated in the front rows of the audience. "Think about that, people. It's not about the lines you speak. What you say comes from what you want. That's what should be uppermost in your mind."

I thought about it. It struck me that my family had had a cherry orchard of its own, and that both its attainment and its loss were with Daisy and me, all the time.

Kingo flung herself back into her seat as if the whole exercise had worn her out and flicked her wrist at them. "All right. Let's try that again."

Most of the women I knew spoke at least two languages, one they produced when they were around men, another they used among themselves. Black women had to master four—one for white men, another for black men, a third for "Miss Ann," and a fourth, their vernacular, for one another's company. But Kingo and Portia talked to everyone just the same—the black cooks at the camp, the mayor of Steamboat, me. How sweet that must be, I thought. How much more energy you'd have if you didn't always have to tailor yourself to the whims of each person you met.

Sometimes Daisy let me watch classes at the Louis Horst studio, which was planted in the middle of a field. There were doors around the perimeter, of warm blond wood, that could be slid open so that the studio was open on all sides. The tin roof above protected the dancers from the sun and the marvelous summer rains that blew cool wind through. The whole building felt as if it had sprung from the forest floor. Ballet fascinated me, with its restraint, discipline, and beauty. Indian dance, taught by a relative of Ravi Shankar's, involved curved fingers and flexed feet. Whatever the dance form, the accompanist played almost inadvertently, watching neither his fingers nor the dancers, his apparent lack of effort mystifying to me. Later, when Daisy cleaned the studio, pushing a dust mop from one end to the other, I sat at the keyboard, trailing my fingers along the chipped keys, registering how casually the players had left their still-lit cigarettes and their half-empty cups to mar the finish.

Then there was the day I heard not piano but the beating of a drum coming from the studio. It was a conga, pulsing out a rhythm I'd never heard before, a beat that made you want to fling yourself forward with abandon. I sat on the platform that ringed the studio and watched. Inside six dancers hung, surged, swayed, and heaved to and fro as if they were daughters of the willows outside. They all had such ugly bare feet and solemn expressions, I wanted to laugh.

The teacher took charge, a small Asian woman in a plain black leotard and tights, her hair pulled back smoothly into a bun.

"And one," she yelled, and leaped into the air with such straight grace, it stopped my breath. When she flew up, toes pointed, arms extended, I forgot what a small woman she was, even that she was a woman. It was as if she had turned herself into an arrow, another entity entirely.

"Haitian uses the whole body, not just the arms and legs," she pronounced, earthbound again. She contracted from the belly as though she had been punched, the surge pushing her backward even as she pulled against it with her arms. Then she repeated the movement, pulling back toward the piano with her spine, but the energy was not in it, and her face was bored. "You can't leave your pelvis back at the cabin."

"Kay, is it back, two three four, or—" The girl contracted from the middle and another invisible blow sent her back, lifting her off the ground until she cut the air with her feet.

I couldn't even feel the muscles Keiko was talking about, but I wanted to move like that too, outside shame or modesty— like a wolf answering the moon.

Most of the time I was at PM I spent with the cooks, who were all black—Peeler, Ruby, Pokie, and America. No matter how evenly Portia and Kingo spoke with them, they tended to keep to themselves.

America Marshall was the exception. I called her Aunt America, in the Southern way. She was the dessert cook when camp was in session, but she also attended to Kingo personally, traveling with her between Steamboat, her winter house in Carmel, California, and her apartment in New York.

The first time I'd met America I was six years old. It was a hot, dusty June day, and Perry Mansfield was preparing to open. Daisy was driving the Jeep from cabin to cabin, rousting mouse families, opening screens, beating mattresses, and mopping the green painted floors. When she finished, she took me to a space below the kitchen at the main lodge, stuffed between a pantry and a laundry room. She knocked on the door and it was opened by a woman the color of gingerbread, with silver hair tamed into a pretty crown of French braids. The room was no bigger than a nun's cell, but the air inside was crisp as a new apple, and the room exuded a cool calm. I just stood back and absorbed it while Daisy launched into the day's plans. A flat of raspberries for Marjorie Perry, Kingo's sister. Strawberries and fresh string beans to the main house for Portia.

America looked at me and smiled. "Lordy, Lordy, what a pretty child."

"Don't tell her that. She's already got a big head," said Daisy.

"You the spittin image of your mama," said America. A flood of warmth went through me. I looked around me. A little pink rag rug was placed just where you'd step out onto it when you got out of bed. House shoes faced the wall, ready for duty. The spotless chenille bedspread had not a wrinkle, and next to the picture of patient Jesus on the nightstand was a small vase of columbines and dandelions—what Daisy would call weeds. None of these things had much impact on its own, but the combination suggested a special care taken with small things, as if America had all the time in the world.

Daisy was always in a rush—to get to the Jeep to get to the garden to get to the kitchen to the freezer to the cellar, the same chaos within and without. I thought of the trash in her yard, the outhouses on the hill, the coal piled on dirt. Her gardens, a survivalist hodgepodge of vegetable and flower, berry bushes and hay, heaped together with no sense of balance or style. And all around the neighbors, mystified by her dramatic, cringing ways. They could not know that every cabbage she grew and every chicken she raised was a banner against the Klan, that the same fear and fury drove her to give them the best of the berries and then to file suit over a fence line.

I recognized America's tidy room for a kind of sanctuary, a place where a person could be at peace. It had never occurred to me that a black person could find that, especially not a person who served. It would never have occurred to Daisy to acquiesce to a position of service, although she required me to clean right along with her from the time I could competently hold a rag. "Don't end up like me, mopping somebody else's floors," she said. That was one admonition I didn't challenge, so America's calm acceptance of her role was confusing. Was she dumb? Was she that most horrible of things imaginable—a nigger?

I was too young to know about the psyche, and the effects of trauma on it, but I intuited that America had not experienced circumstances like Daisy's. She too had lived in Nebraska, but I knew little else about her. Had she ever seen a family member murdered, or experienced lynching as a daily possibility? I couldn't have articulated her effect on me, but over the dozen or so years I knew her, she impressed upon me an alternate example of how a black person could be. Even when she scolded me for stealing a warm cookie from the rack in the dessert kitchen, she did it gently, without even turning around. "I know you don't think you can slip up and steal and me standin right here in this room," she'd say, as if to herself. "Yes, Jesus, that chile don't believe that. You been helping her right along, so she wouldn't even dream of trying to be sly. Thank you, Jesus. You always know the way."

I went to stay with her in town at the end of one summer, when she was packing up Kingo's house for the winter. Her room above the rest of the house had a sparkling indoor bathroom with mirrors and lights, plenty of fluffy towels, heat that came out of mysterious holes in the wall, and the same characteristic tidiness. I had to touch everything—the wallpaper, the deep rugs, the window latches with their fancy handles. And downstairs, so many fireplaces—and a violin. America told me Kingo played with the Monterey Symphony. She walked among the beautiful, expensive things as self-possessed as ever, without an ounce of covetousness.

She made us breakfast, washing each berry as if it were the first and only one. Then she took fresh yogurt from the oven, where she'd made it in baby food jars, and set out wheat germ. I sat down and prepared to pile in when I realized she was about to ask a blessing. We sat a moment in quiet, mine restless and anxious, hers peaceful and still, before we began. I had been taught to clean my plate, and I did, although I didn't like the yogurt or the wheat germ, and I searched in vain for sugar. But the strawberries, which I ate all the time at home, I tasted as if for the first time. Without my knowing it, America was awakening in me the distinction between poverty and simplicity.

I was amazed by what was not in her. Like Daisy, I had believed that if you took away their offal, you were forever doomed to the shadows, like a chambermaid who spirits the chamber pot out of the room or a prostitute who scurries about in the dark, servicing the gentleman. But America was not tortured by the prospect of one more day spent serving others. Washing Kingo's panties—by hand no less—did not disgust or oppress her. It seemed to me that whites didn't have bodies, and blacks were only body. Daisy couldn't have articulated that dichotomy, but she railed against it every waking moment— not because she disagreed with the fundamental premise, but because she wanted to be a lady too. She did not want to be relegated to the shadows.

Daisy and the cooks declined most of the invitations they got to performances at PM, which were readily extended. When they did attend, they seemed shy and uncomfortable. Sometimes, if Daisy was in a good mood and I promised her the moon, she let me go. Once Portia Brown, the daughter of the great dancer Harriette Ann Gray, took me to a performance her mother had choreographed. And once in a while, Daisy got permission for me to attend a theater or dance class. In one modern dance class, we attempted to inhabit the spirit of our favorite animal guides. I didn't understand what made this "modern," but the choice was easy: I was the agile weasel who clothes herself in ermine come the cold.

More meaningful to me at this age was the chance to join in the occasional campfire sing-along. At one, I learned marvelous naughty songs from the other kids—*Oh she stood up there in the midnight air. And the wind blew up her nighty. Her boobs hung loose like balls on a moose and she thought she was God Almighty.*

Maybe it was that evening—or maybe it was another night—that I was coming down the path in the gentle evening air, dusk settling quietly and the night birds waking for their evening hunt. I was luxuriating in the sense of feeling, for one brief hour, unashamed, unself-conscious, normal. We had sung:

L-O-double L-I. That's the only decent kind of candy. (Candy.)
Man who made it must have been a dandy. (Dandy.)
L-O-double L-I. P-O-P you see.
It's a lick on a stick guaranteed to make ya sick. It's a lollipop for me.

I was belting it out so intently that I didn't see Daisy waiting for me with her mop and bucket. When I did spot her, I almost gave in to the impulse to hug her, I was so happy. I didn't notice how she was standing, practically coiled. And so it came as a particular surprise when she spit in my face—out of nowhere a spray like hot buckshot, acid with loathing, spewed with such force that it stung my eyes. She didn't move then, as if she was as surprised as I was. We just hovered there in the dimming light, the trees watching. Then she turned and said, "Come on," and headed for the Jeep. She put the mop bucket in the back and climbed in next to me, and all the way home I could smell the stink of her dentures on me.

~ ~ ~

At the end of the summer, Perry Mansfield vanished from my life. Plywood was nailed over the cabin screens, the gates were locked, and the entire place went into hibernation. And along with the camp, art vanished too—almost. I had a secret place where I stored it: in the root cellar Ernest had built, along with the beets and carrots and potatoes nestled in the cool pale sand that was supposed to keep them cold enough that their sugars would deepen, and warm enough that the winter cold wouldn't nip them.

It could have been a dank place, but Ernest had built it perfectly. He'd dug a hole nearly eight feet deep and braced it with poles, the kind of hole the Apaches used to dig when the cavalry was hunting for them—or so Ernest said Mr. Anderson had told him. The top of the mound was almost completely obscured by pine branches and dirt.

It was my chore to go fetch a mess of potatoes and carrots from the cellar when Daisy was making a roast. At first the cellar scared me, but then I realized it could be my secret room. It was surprisingly warm down there, even on the coldest nights. Little by little, I started keeping stuff there, in one of the crocks—Gampa's cigar box, a pack of Ernest's Pall Malls, a sepia-toned magazine I'd found, about black ladies and their gentlemen callers. Now I added the ragged leotard that still smelled of toe chalk and the toe shoes Daisy and I had salvaged from the trash. Before long they would take on the scent of their roommates—carrots starting to curl, rubbery parsnips, nearly black beets, a peck of apples.

Sometimes I untied my Buster Browns and put on the toe shoes, which were way too big. They were a conundrum, soft and yet iron stiff, the bituminously shiny black satin laid upon the hard toe. I made certain to stay on the clean planks that Ernest had laid down along the shelving. I pulled my spine up like Keoko would have. I knew the ballet positions now—the arms, the feet. An imaginary string at the nape of my neck connected to the roof of the cellar, and the outside muscles of my thighs pulled into a tight fifth. *And one.* My arms drifted into the perfect oval of first, rooted from the back like wings. I pliéd, pulling up even as my legs went down. Holding on to the beet crock, I willed myself into a deeper turnout, brought my arms up like a crane ready to leap with perfect faith into the winter air.

5 ~ Quilt

Gama with Rose Marie and Mary

~~~~ **When I woke,** the back of the house seemed brighter, the bird chatter louder, more cheerful. Once more winter was on the run. The frost that used to fill the entire window in starbursts had diminished to a few lacy wisps at the bottom. Still, I snuggled under Gama's quilt like I always did before it was time to get up. I ran my hands over all the patches, little buds of love and sadness blooming. At nine my life was like a page of

paper dolls where most of the others had been popped out, but here in Gama's quilt all the absences were present.

This was her last quilt—a winter one of heavy cloth fitted over several layers of batting, mostly in the dun shades of a late-season forest. She'd made it the summer before she died. In the patches, the hoarded remnants of the departed—Mama and Daddy, Rose Marie and Mary—mingled with those of Gampa, Daisy, and me. Gama, the freshest loss, scarcely bearable, still held us all together with her careful stitching.

Gama's favorite time to sew was the morning, when the house emptied of everyone but us and the light spilled in softly like rain. Gama and I would sit together by the kitchen table that faced the meadow, while Jiggs supervised the yard outside on the doorstep. Just outside the window, new tendrils curled on the sweet peas she'd planted. Behind us the stove wound down, but she had the coals banked for Gampa's dinner, which he'd expect, as he had for the sixty years they'd been married, promptly at noon.

She'd wipe the oilcloth and lay out her fabric scraps. She'd move the pieces around as if she were imagining the finished look, then she'd take out the small gold scissors she kept in her sewing box and nip away at edges until blocks, swathes, and rectangles emerged. She'd baste the pieces together into blocks until she was ready for the final stitching with her good thread. Sometimes, as she worked, I made a dress out of scraps for my doll. We could spend hours that way without speaking, the only sounds in the room the scissors and the rustle of fabric, except when the peace was broken by the geese reprimanding Gampa for moving the hose too near their pen, and Jiggs in turn scolding the geese.

The patches, never exactly symmetrical, held together companionably as Gama transformed them into a circular diamond pattern she called "turkey in the straw." Her hands moved rap-idly, long fingers gracefully gathering the pieces, then marrying them with straight pins stashed in the corner of her mouth. Her stitches, fine and even, hardly showed against the dark blocks. When she ran out of black thread along the lower edges, she al-lowed herself the extravagance of green. After the final sewing, she sprinkled the seams with water and pressed them open on the wrong side with the big flat iron she kept hot on the stove. When she laid out the finished block, I marveled at how snugly all the pieces fit: Mary's gray skirt to Mama's rough tweed jacket, Gampa's brown corduroy pants to the sleeve of Daisy's overcoat, the rust wool coat I had outgrown to a storm-blue swatch from my father's work shirt, the only thing of his I ever owned.

Even Rose Marie was here, in a swatch of red velvet stitched into the edge of the quilt, as if she'd nearly slipped away but had been flushed out and caught. Nobody told me that the scarlet that popped among the patches of gray, brown, black, and ma-roon was hers. I just knew. I had studied the picture of Rose Marie at my mother's funeral, in the old black album where pho-tos of happy fishermen and trout-laden twigs were interspersed with pictures of Mama in her casket of snowy silk. Daisy kept the album in the buffet, which seemed to be the repository of memories of the dead. Sometimes I looked at the other things stashed there when she was gone—Mama's death certificate that looked like a diploma, the books about Robert Ball Anderson and Geronimo, and the gunnysack Ernest used to drown "var-mints," in which Jiggs's mother, brothers, and sisters had surely gone to their deaths.

Rose Marie was not dead, far from it. But she was long gone, and good riddance too, according to Daisy. "Got herself a reputation," Daisy sniffed. I had no idea how you earned one, but I figured it had to be some kind of mischief to get this kind of rise out of Daisy. "Had the whole town talking. It's a damn shame." When I said that I never heard that from anyone else, Daisy's eyes turned harder. "Nobody would tell you anyhow. But you'll find out soon enough." By now I recognized her tone as the one she only used when she was speaking about women "going with" men. How fascinating, I thought. My sister must have done something with men so scandalous it couldn't even be spoken.

Rose Marie had gone to live in California, the place where they'd invented glamour. I no longer prayed "Hollywood be thy name," but I knew that Hollywood was where movies were made, and that it was not only legal to be pretty and sexy there, they wouldn't let you in if you weren't. It was a place where Marilyn Monroe let the wind blow her skirt up, and Jane wore a leopard-skin dress that revealed the tops of her bosoms to Tarzan, and no one called them "fast."

"How could she go up Fish Creek Falls and let them see her?" Daisy would fume, as if Rose Marie had only done it last week. "Does she think they won't talk about a black girl?" See her *what*, I wondered, but Mary said only, "She didn't care."

Wow, I thought, how cool to do what you wanted without being paralyzed by what this white man believed or what that white woman might say, until every move you might make froze inside and all you did was stare out the window longing for what might have been.

"She ought to be ashamed!" Once Daisy got going, she couldn't let it go. Why Rose Marie shouldn't have enjoyed herself, I couldn't figure out, but I sensed it was legal only for men to do that. There was something nuclear about being black, female, and pretty. I imagined Rose Marie in matching peach high heels, handbag, and gloves, smiling out from under a rakish hat with one long feather as if she had a secret. She might have left a plume of scandal in her wake, but she was living proof that you could go against Daisy and the whole town and survive.

Gama's quilt had a whole big swirl out of that charcoal gray skirt of Mary's. It still felt as soft as a new lamb's wool, and I had nearly worn it threadbare, resting my cheek on it when I went to sleep. How smoothly it had sheathed the swell of Mary's hips and the length of her legs that Sunday Danny Reeves came calling.

Mary had taken special care dressing that day, dabbing Evening in Paris behind her ears, carefully sliding on the silk stockings, making certain the seams were straight. Then she said, "Go in the kitchen. You don't need to see where I keep my things." I already knew that she stashed them on the shelf above the tablecloths, hidden from mice and moths and me. Nothing pleased me more than to inhabit her shoes and clothes, fingernail polish and lipstick. I slipped through the curtain that separated her alcove from the rest of the house, then inched my way back gradually, unable to resist the spectacle of Mary making herself beautiful. Maybe one day I could inhabit her grace and mystery and make them my own.

She smoothed the skirt and tucked the short-sleeved pink sweater just so, tying a little white scarf around her long neck with the ends arranged cunningly to one side, as if to show off her carefree ways.

"Now I want you to get out and stay out when Danny comes," she told me. "Do you hear?"

She unpinned the spit curls in front of her seashell ears and smoothed out her eyebrows before the mirror, double-checking her teeth and then bringing out her lipstick, Fire Engine Red by Revlon. Careful as a painter, she outlined her mouth before filling in her lips until they looked like a big bow, more perfect even than Elizabeth Taylor's.

Danny was the Trailways driver who'd driven me and Mary back from Denver the night before my mother died. He was so tall that I worried about his head when he came through the door, dressed in a checked shirt and trousers. I had never seen him without his uniform, and was surprised to discover that beneath his cap, his hair was the color of a field mouse. But his eyes were direct, confident, and blue as the Colorado summer sky, and he brought the reassurance of a capable man into our house of women.

"Wow," he said when he saw Mary, and I could feel her warm to the knowledge that he thought she was pretty.

I knew I was supposed to go and leave them alone, but Danny reached out and pulled on one of my braids. "So how's it goin, sweet pea?" Tentatively, I sat down on the edge of the couch.

Behind him, Mary's eyes hardened. "Scram, pip-squeak." Her soft laugh did not conceal how thoroughly she meant it. Savoring the smell of Old Spice on my hair, and knowing I'd pay for it later, I shuffled through the curtain, but not before I looked back and saw that Mary had settled herself by Danny again and folded her beautiful hands primly in her lap. He had eyes only for her then, and no wonder.

But Mary hadn't married Danny. Instead she'd gotten the job at the radio station in Craig. Daisy marveled about her day and night—the very idea of someone from our family working in an office. And then she got accepted to Loretta Heights, a Catholic college in Denver. "I believe she'll do it," Daisy kept saying in wonder, as if by enrolling Mary would assume the burden of redeeming all the family's losses, failures, and catastrophic bad judgments. It did not surprise me when we didn't see her much after that.

It was a summer night a couple of years later that Jiggs went crazy, snarling and barking his head off. Then the geese commenced honking so loud I just knew a fox had slipped under the fence to steal the babies. Daisy jumped out of bed, grabbed the flashlight, and headed to the front of the house to peer out the window. The chain clanked against the front gate—no fox would be fiddling with that. Only then did she turn on the yard light and see a man hesitate before Jiggs, who was jumping up and down on his hind feet, making it clear he wanted to take the visitor's face off.

Daisy opened the door just enough to be heard. "Who is that sneaking around in my yard in the middle of the night?" she called. Whatever the response might have been, we couldn't hear for the racket coming from the geese, as they joined Jiggs in his protest.

"Rita, didn't I padlock that gate?" I was certain she had. Daisy padlocked everything—the shed, the dog, the well house, the cellar, the tools, the berry bench, and, of course, the house.

The man eased himself between the goose pen and the Jeep. Bess honked, put her head down low, and ran at him, hissing. When he didn't back off, she hurried to huddle over her goslings, wings extended.

"I said, who is that?" Daisy repeated. "Don't make me have to call the law." She reached behind the buffet and brought out the .22 rifle.

"If you're Aunt Daisy, I'm your new son-in-law," the man hollered up.

Daisy didn't say anything, but I could hear her gasp. "That's bullshit," she said under her breath, not loud enough for him to hear. We could feel him waiting. I had never seen Jiggs so fierce, showing all his teeth, the fur all along his back bristling high.

"Can your dog get a-loose?"

"How'd you know to come out here?" Quietly Daisy double-checked that there was a bullet in the chamber and then slid the bolt action into place.

"Mary told me all about you and Rita Ann and all." This time we both caught our breath at the sound of my name. "I just thought I'd come out and get acquainted." He edged closer up the drive past the Jeep. Jiggs's barking gave way to a menacing low growl, and he lunged to the end of his chain, snarling and choking.

At last the man paused, at the bottom steps of the house. He lit a match with his fingernail and brought it up to light a cigarette, giving me my first clear glimpse of his face. With his threadbare cap and coveralls, he looked like a Steamboat boy who wasn't going anywhere, the kind of white guy who might sell us a new radiator cap at the auto parts store.

"Why you wait till it's black dark to come out here?" Daisy asked.

That brought a laugh, as if he was bashful. He kept looking around the yard like he was searching for something he didn't see.

"I honest to Goshen don't mean no harm. I was just hauling down Highway Forty and thought I'd drop in to say hello. My name's Lonnie."

Daisy looked at me. Nobody went out of their way to climb over all three passes to Steamboat Springs unless they meant to.

"So, how long you two been married?" Daisy asked, in a flat voice.

"Couple three, four months or so. I ain't too sure." When he smiled, it didn't extend all the way to his eyes.

"You don't say," Daisy said.

Lonnie inched closer, put his foot on the bottom stair.

"Mr. Lonnie Whatever-Your-Last-Name-Is, I don't mind shooting you, but I don't want the mess. Now you just stay right where you are and we'll talk."

Lonnie looked up sharply, his eyes widening, as if he was trying to see if Daisy actually had a gun. I edged around her to see past the geranium pots. "Stay out of sight," she whispered fiercely to me. "No tellin what this fool wants." She turned back to the window. "Exactly what become of Mary Loretta?"

"Oh, she took and went on back to Indiana, visitin my folks."

So this unexpected brother-in-law of mine was a Southern white man. In spite of how sneaky he seemed, I was curious. Maybe they'd let me come visit.

"You not gonna let me come in at all?" he added after a pause, as if it had finally dawned on him that Daisy wasn't exactly welcoming.

"What do you do, Lonnie?" The way Daisy said his name, I knew he wasn't getting in our house anytime, ever. In fact, he'd be lucky if she didn't shoot him right there, and then call the sheriff to come clean up the yard.

"Oh, you know, one thing and another. I'm a jack of all trades." He put one foot up on the chopping block, crossed his arms, and leaned on his leg. "I really would be happy to help you all out around here, if you need some work done. I got a fine set of tools." He smiled, looking around at the rows of cabbages and onions in the patch behind him. "Sure is a pretty place."

"Where you two plan to set up housekeeping?"

"Well, maybe we'll go back to Indiana and maybe we'll go someplace else. It all depends."

Something about that answer shifted all of Daisy's alarm into something else. Whatever it was, I knew she was done.

"It's time for you to go, Lonnie," she said, more softly now, almost nicely. And Lonnie stopped smiling at last.

"I'm sorry for getting you-all up out of bed," he said. He pushed his cap back and ground out his cigarette in the dirt.

"Don't mention it," Daisy said. And although she only watched coldly as he backed down the driveway toward the gate, I had the sense that he was still hoping she'd invite him in. Then the gander honked, and that set Jiggs off again, and Lonnie, flustered, almost tripped over a ring of stones flanking the hollyhocks. Embarrassed, he pulled himself up taller than he really was and strode out the gate.

Daisy didn't put the rifle back for a long time. She just stood there watching the red taillights of his car go down the hill and then cross over toward the road that would take him back to town. Then she said, "All along, she was just biding her time, wasn't she?"

And after she turned out the light and we were back in bed, she said, "Don't never take nobody else's child to raise, Rita. You'll be sorry if you do."

Next morning she did something I had never seen her do before. She kept her face to the wall and didn't get up all day. Not to drink or eat, not even to pee. Not even to water the cabbages, which would turn bitter if their roots didn't stay wet, or cover the strawberries with cheesecloth so the magpies couldn't get to them.

~ ~ ~

In the weeks and months after Mary left, I'd stand in front of the mirror and talk to my lost family, struggling to reenact the life I thought they'd lived in Nebraska when they were rich, before I was born. I planned tea parties with Mary, costumed patient Jiggs for rodeos and roundups, saddled the whole family on herds of mustangs and Appaloosas, my father on a quarter horse mare with a sure foot and a sensitive mouth, Mama on a paint.

It seemed to me that Daisy grew harsher then about the family, casually grinding down my versions of them all without considering that I still needed to believe in them, no matter what they had been to her. "I tell you this, back in Nebraska, we had everything, but your daddy didn't care about nothin but pickin that damn guitar. Mr. Anderson thought the world of your mama. She could have been somethin if it hadn't been for him—in the end all she come up with was a bunch of damn babies." Daisy no longer mentioned Mary at all.

I seemed to have arrived just when everyone else was headed out of town. I'd been told that Daddy had come back briefly after my mother's death to try to assert custody over me, but Daisy had easily prevailed—fueled more by spite against him than attachment to me. The only ones left were those who couldn't get away—Daisy, Gama and Gampa, who, already in their eighties, were too old to go anywhere else. Ten times eight—a sum incomprehensible to me, and Daisy nearly sixty. The only concerns they seemed to have were the welfare of poultry, how soon the steers would put on fat, the best way to grow Irish potatoes, whether we should limit our bees to raspberry blossoms to make the honey sweeter.

I felt the unspoken rhythm that bound my grandparents— Gama dipping up water for Gampa even before he came into view—but I didn't love him as I did her. "Heifer" was as strong a rebuke as I ever got from her. Gampa never smiled. His main concerns were "can't" and "don't."

"Put that saw down. Girls can't handle saws."

"Don't ax so many damn questions. Shit a mile—you run me crazy."

When he came up to the house after watering, his wide hat

smelled of damp straw, and only his suspenders held his
trousers to his gaunt frame. His joints, worn down to bone on
bone, hurt him all the time, and he walked with a cane. I was
sorry about that, but he wasn't much fun. Like Daisy, he was always carrying on about the South. I was so tired of the Civil
War, slavery, the trials of the sharecroppers. Sorry old tales
from sorry old days. I sipped lidfuls from his bottle of Mogen
David wine, sweet as sorghum and dark as ink. It made me
woozy, with its slightly rotten smell, but every day I had some.
It seemed to make things go a little better. "You gon' miss me
when I'm gone," Gampa said to me, over and over. I never said
anything, but I wondered, *Why?*

Gama had used his brown corduroy pants to ring the outside of the quilt. I remembered how the seat had worn slick, and
finally they'd begun to smell of old man's pee. Gama had filled
a dishpan with water and shaved a sliver off the block of homemade brown soap into it. But after she soaked the pants for
days and they still smelled, she gave up trying to save them and
chopped off the legs for her yardage basket, discarding the rest.
Grandpa would hunt for them for months—he was always
looking for something he'd lost. "Dang it. Who done made off
with my galluses?" His suspenders had a life all their own. He
seemed to me like an old oak transplanted from that South he
was always talking about, safe now, yet bewildered among the
aspens, and in need of warmer weather.

Yet it was Gama, not he, who left me next.

December always made it seem as if the holiday season had
been invented especially for Steamboat Springs. The abundant
snow was fine as down, making the Christmas lights brighter,

the carols merrier, the decorated trees in the front windows almost unbearably picturesque and suggestive of family cheer.

I could hardly stand it. Five winters on, Mama's death still colored the season. Daddy had sent me ten dollars, the money tucked into a card decorated with a frosty crust on a jubilant Santa and signed by my stepmother, Marge. When Daisy said that was because my father couldn't write—"Ain't got but a third-grade education"—I felt a stone bruise my heart. I was in the third grade, and I could write. Why couldn't he? I wondered why I hadn't been worth fighting harder for, how he could be so ineffectual when Daisy was so strong that she could mow me down with a glance and not even know she had done it. Was I like him?

Daisy usually forced visitors to stand in the yard while she stood above them on the steps. The prospect of carolers expecting to come inside and warm up after they sang filled her with distrust. She pictured them as singing thieves. "What they're doing is trying to see what we got," she fretted. I started to point out that nobody would be interested in what we had, but thought better of it.

Two days before Christmas, however, a lady from one of the churches arrived at the house with donations for "the poor," and there was no way Daisy could refuse to invite her in, struggling on the porch with a box so heavy it looked like it would break apart. The climate of the house shifted as she lugged it across the threshold. I could feel it in the way she and Daisy ramped up the panicky pitch of their good cheer. It was in the set of the woman's mouth as she looked for a clear space to put the box down, taking in the jumbled room, the smell of kerosene, the mud tracks on the floor.

I had been to this lady's house. Daisy cleaned for her. They had special rooms. They ate their breakfast in a nook. They called their supper "dinner" and their dinner "lunch." They washed their clothes in the laundry room. Each one slept alone in a separate room. I had no idea what they did in the one they called the living room. We did everything all in the one.

The truth was, I was starting to see how different we were. We bunked among our belongings like cowpokes, with no regard at all for how they looked. We didn't have anything pretty—no pictures, figurines, or doilies. Daisy's reading chair found its slot in front of the washing machine across from the Deepfreeze. My piano abutted the wall across from the built-in ironing board. When Gama's mahogany buffet seemed too tall, Daisy got out the saw and shortened its legs, not even bothering to smooth the edges. This did not mean Daisy was unaware of how other people would judge our household. And it certainly didn't mean she was unconcerned—she just seemed at a loss as to how to do it differently. Watching this woman viewing us as "the poor" riled me, though. But she and Daisy had apparently already found common ground. "I got to be her mama and papa both. Don't nobody give me a dime." It sounded like something Hank Williams would sing.

The box was full of groceries—canned goods and oranges and fruitcake, which made no sense to me at all, with its heavy candied rinds and stink of ferment. I relished the bag of nuts in their hulls, though, and the thought of the feast we would have when Daisy settled in with hammer and nail.

She turned to me. "Tell her thank you, Rita. Give her a kiss."

I looked at the floor. I felt like Jiggs, being ordered to give his paw to a stranger.

This little display of disobedience appalled Daisy. "Did you hear me talking to you?" she said.

"Thanks," I said, without approaching the woman.

Daisy was about to drag me over to her when the woman mercifully intervened.

"I got another box in the car, Daisy, and I want to get two more sets delivered before dark."

The second box didn't seem worth much at first, mostly old clothes with lived-in smells that were way too big for me. But then the woman said, "I brought you some books," and reached deeper into the box. *The Black Stallion's Blood Bay Colt*, *The Black Stallion and Satan*, and *The Black Stallion Returns*. I let out a cry.

"She's crazy 'bout a book," Daisy said. "I've got to hide the flashlight under my pillow so she won't use up all the batteries reading under the covers."

Now in the evenings, I raced along, rapt, as Alec Ramsey stole the heart of the Black Stallion and finally was able to ride this horse that no one else could tame. I didn't even mind so much that school was closed through New Year's and I didn't have the company of my friends to alleviate Daisy's oppression. Daisy was the kind to break a horse with a two-by-four, but this horse could stand up to even her.

But as if to remind me of how precarious any tranquillity was, the weather turned treacherously unstable. A chinook blew through, its warm winds convincing herds of cattle on the eastern slope that spring had come and coaxing them to lie down and go to sleep. That night a blizzard swept in. When morning came, they found themselves frozen to the ground. The hard top crust collapsed under those who managed to make it to their feet, breaking their legs. Trees actually exploded in the sudden temperature drop. I wondered at the arbitrariness of life and death. Was this why people called nature Mother and God Father?

Then my own personal chinook blew through. I developed a raging throat infection, which Daisy had insisted on treating with hot toddies instead of taking me to the doctor. My temperature soared until it felt as if my brow would burst into boils. Each night Daisy stewed up rye whiskey, vinegar, and sugar, which scalded my raw tonsils but healed nothing. In the back of the house, I floated aloft on the fever for what seemed like weeks, until Daisy got scared enough to call Dr. Smith. Then in a panic she got down on her hands and knees to scrub the floor, terrified at the thought of him seeing our messy house, all the while invoking Jesus. "Now the kid's trying to die," she moaned. "Lord, why you do me this-a-way?"

By the time I was back on my feet and back to school, the January cold had gotten down to business, as if to punish us for too much holiday foolishness. Even Jiggs with his thick shepherd coat stayed curled up under the steps with his tail across his nose, too cold to greet me. And something was wrong with Gama—something so bad that I came home from school one day to find the twin beds Daisy and I slept in hauled off to the basement and Gama moved to the back of the house.

"Mama's going to stay with us from now on," Daisy said.

"What can I do?" I asked.

"Move your stuff down to the basement and keep your head in them books."

I liked having Gama close, but I didn't care much for sleeping in the basement, with the coal bin at the head of my cot, and the big potbellied stove with its iron skirt, even with Gama's quilt to keep me warm. Daisy had kept me down there for two weeks when I had chicken pox—to keep me from going blind, she said. The concrete on the walls had bubbled because Daddy had let it freeze when it was setting up. He was the reason the foundation leaked in the spring too, letting the melt puddle the floor. Daisy always looked at the walls and grumbled, "Don't never get a man prettier than you. They no count." I ran my fingers over the rough surface and wondered whether, if I died down here, Daddy would come back to mourn over me.

Now in the evenings the house was way too quiet, the rattle of the newspaper too loud as Daisy turned the pages. I read on with a vengeance, as if by opening my book I was climbing into the stall of the wild black horse who would not bow in the face of fear.

One afternoon Daisy pulled a package of antelope from the freezer and said, "We ain't got but one package of elk steak left, and I'm saving that till Ernest comes." We were down to boiled ribs for supper again, but at least Gama could have broth from the marrow.

The house creaked, and the melting icicles hanging thick as swords from the eaves stopped dripping. I coiled myself inside the shadow of the Black Stallion. I could feel his soft muzzle, hear the rhythm of his hooves pounding soft turf, imagine how he smelled after a hard run. Gama's sickness was rising until it filled the whole house, and the big black horse was all that stood between me and it.

From the table, Daisy and I could hear Gama fighting to breathe. I struggled to think what to do for her. Each time I went back there the bones had risen even closer to the surface of her thin skin. She seemed to be evaporating before my very eyes. I felt myself sinking beneath an old feeling, the dread that once more I was about to lose someone I couldn't live without.

Daisy rattled the paper, one twig of a braid peeking out of her bandanna. "I reckon it's time for somebody else to die. Death does come in threes." Was she talking about me? Gama? She pointed to a picture of a car wreck on the front page of the paper. "You see that, don't you? He got what he had coming.

READ

Always drivin too fast. Nearly runned me off the road last winter." She read every corner of the *Pilot,* but she started with the obituaries.

"You can always count on your regular crop of dead youngsters," Daisy said, settling in for a cautionary litany. "Seems like they all ripen and die in a clump." She counted them off on her left hand as she propped the paper against her lap with the other. "One'll drown in the lake. Another'n'll go simple and let his horse ride him into a tree limb." I looked down at the cover of the book in my lap, traced with my eyes the wild arch of the stallion's neck. Quarter horses' necks were nothing like that.

Daisy's little finger popped up—"hunting accidents." She looked at me over the rim of her glasses. "Hardly ever fishing problems. That's why it's better to guide folks to fish instead of fooling with twelve head of horses and Texans just dying to shoot themselves all to pieces."

The bedsprings squeaked behind us and Gama called out, "Oh Lordy, I'm so cold. Mother, is that you?"

I knew Gama had a mother, but I could hardly imagine back that far.

Daisy jumped up, dropping the paper. She bent down to adjust the grate in the floor to let up more heat from the basement, even though the house was already so hot that sweat dribbled down between my shoulder blades. She went back to check on Gama.

"Mama," she said, "quit kicking off the quilt." I could hear the slosh of water in a cup. "And something else. You got to drink enough to take your medicine because I just caint stand no more tragedies." Gama's throat clicked as she took a swallow and then she let out a terrible moan.

I dived back into the book. The shipwreck and then the

Black Stallion swimming to the island. *Oh hear us when we cry to thee, for those in peril on the sea.*

Daisy back, resuming her conversation with the newspaper. "Like I was tellin you, tragedies come in a clump of three. Father, Son, and Holy Ghost. Like last fall when that eighteen-wheeler carrying them steers burned out his brakes on Rabbit Ears. Next it was that bunch of teenagers hopped up on boiler-makers in their folks' pickup all mashed to hamburger with the hot radiator landing in their laps. And farm machinery just loves to roll over on a fella. It's a damn shame."

I walked to the edge of the piano and looked back at Gama. Her hands lay motionless on top of the plain gray quilt, as if frostbitten. I had never seen them so still. The Cherokee and African bones stood out in her beautiful face, with its cantilevered cheekbones and wide mouth. She opened her eyes and looked at me as if she were seeing me for the first time, or the last. There was a smell to her now. I was afraid to approach her, and ashamed of my fear.

"Baby?"

Behind my back, I clutched the book so tightly I could hear the spine crack.

"Don't look so puny, else when the monarchs come, I'm gonna grab the first one I see and make you eat it. Give you some spunk."

"Gama, what if I brought you a present?" I had no idea what. I didn't even know if she heard me before she nodded off again, as if she were sliding down beneath the weight of her pain.

This time not even the doctors were capable of beating back disaster. Gama had cancer of the throat, and it was beyond the point of treatment. Daisy insisted on caring for her at home, so the quiet, anxious evenings continued for a time.

It seemed she evaporated much faster then. Soon she could neither eat nor speak. She grew paler, as if the tubes that were supposed to drip fluid into her veins had somehow reversed course and begun to drain her. I wanted so badly to give her something that would make a difference. I reviewed my stash down in the root cellar—a coloring book, my horse books, a couple of cat's-eye marbles, my doll. Nothing seemed likely to be of much help.

One day when we were in town after school, Daisy sent me into Rexall's to pick up a prescription for Gama. The drugstore had black and white tile floors and smelled of hot water bottles and liniments. A cardboard ad of a woman with organized yellow hair like pulled taffy caught my eye. Smiling, she held out four pieces of fancy chocolate nestled in crinkly paper tubs. The box of candy looked like the key to a world where fathers always stayed and mothers never shrieked and nothing ever smelled bad. It cost $1.27. I had three quarters. As I stood there, snow melting off my rubber overshoes, I calculated that if I saved my allowance for two more weeks, I could buy one for Gama. In fact, if I volunteered to help Daisy clean the church on Saturday, I could earn a whole dollar. Although if Jesus, who was white and had a powerful father, couldn't figure a way out of this thing called death, how could Gama?

The next week I was back, cash in hand. But the cardboard candy lady was gone, and the stack of candy boxes with her. I went to the druggist.

"Do you have any more White Man samplers?"

He peered at me over his glasses. "What?"

"Those chocolates in the yellow box."

"It's Whitman," the clerk said, smiling as if I'd said something cute. "They're in the back now."

Whitman. That surely couldn't be right, I thought, but then he brought one out. With great relief, I pulled off my mittens and gave him the wadded dollar bill and change.

It was only five miles to our house, but in the early winter dark, it was hard to know where the shoulder of the road lay. The fields, full of pillowlike mounds of snow, swallowed sound, and the horses, huddled close to their hay sheds, wore shaggy beards of frost. For once, Daisy, preoccupied with keeping the Jeep on the road, was silent. I clutched the box of candy in my pocket, praying that Gama would still be awake when we got home.

When we pulled into the driveway, the dark house looked abandoned under its fringe of icicles. Inside, the house was so cold that the ribs Daisy had set out to thaw that afternoon were still frozen solid. And the light was gray, as if it had been bent back on itself by cold.

"Damn it, I banked that fire good. Wonder what made it go out?" Daisy said, rushing to the stove. She raked the ashes out of the bottom and poured kerosene on the kindling to hurry the fire. I went back to Gama with my box of candy. The shades were drawn and a half-empty bottle of Vicks on the sewing machine filled the room with the smell of eucalyptus.

"Gama, I brought you a present," I whispered. At first she didn't move. Then she swam up out of herself and looked around the room as though she was startled to find herself still there. Her eyes landed on me, and after a moment her hand reached out to me.

I touched her forehead. Her face was feverish in the cold room, as if she had been possessed by a spirit that surpassed the winter and the cancer, making her strangely strong. I showed her the box. She tried to hold it, but it fell back on the bed.

When I pried off the ribbon and removed the lid, she smiled through dry lips.

"Now ain't this'n a play-pretty," she said hoarsely. Then she frowned. "Rita, you getting your lessons? You know you got to get them lessons." It was the same thing she always said. But she didn't touch the chocolate.

I scanned the lid of the box. "Gama, how about this one? They call it a cherry cordial. Something like what you would mix in a fruitcake."

"In a bit, baby," she whispered. "I don't feel so hungry right now. Why don't you taste one for me and tell me how it is?"

"But Gama, I bought these for you out of my own money," I said. Maybe if I pushed, she would just snap out of it.

She sank deeper into the pillow and closed her eyes.

I couldn't let it go. "Gama, you have to at least try it." Even I liked the cherry ones, though I didn't care for chocolate at all. I took it out of its wrapper and put it to her mouth. Finally, she took it between her lips and tried to chew. Then she coughed, and began to choke, struggling.

"Daisy," I screamed.

Daisy was just coming in the door with an armful of wood. She dropped it and came running.

"Mama? Oh Jesus, she's throwing up blood." Then Daisy saw me with the pleated wrapper in my hand. "Is that candy in her mouth?" She hit Gama on the back, dislodging the cherry. Then she turned on me. "Jesus, Rita. Don't you know she ain't got no throat?"

How could I have been so stupid? Shame and sorrow swept over me.

"Get downstairs," Daisy said, hurrying toward the phone.

I went down into the damp, dark basement and huddled under Gama's quilt, pulling it up over my head to block out the draft coming off the cold cement wall—to block anything that would make me feel my body. Maybe, I thought, I could kill myself, I could die. But I didn't know exactly how to do it, and I was afraid of pain.

The white-coated men came again, with their low hard voices. Overhead, the floor creaked as they walked back from her room. Finally the night claimed all sound, and everything fell deeply still and dark, even the fire.

Next morning, holding the quilt around me, I climbed the stairs carefully, avoiding the boards that creaked. But I could feel that something had changed in the house. And when I went to the back of the house, I knew, even before I reached the bed, that Gama was gone.

6 ~ God's Goose

Me at twelve

~~~~~ **"Time come you** need to learn a lesson."

It was a Saturday morning, buttery sunshine outside making the snow sparkle so that its crystals gleamed like flashing lizard eyes. But the atmosphere in the house remained charged from the night before, when Daisy had found a *True Romance* magazine in my sock drawer. "Where did you get this?" she'd said, flaring. I'd shrugged and looked at the floor.

Now she stood before me in her pajamas, full of mysterious

resolve. Never mind that we hadn't even had breakfast yet and Father Funk expected us to be at church to clean by nine.

"Get downstairs," Daisy said.

"Why?" I asked, apprehensive. It had been a few years since Gama had passed, but I still dreaded that place I associated with illness and death.

"Git," she said. "No sassin."

I went down the stairs and sat on my old cot. The slop jars were stashed underneath, smelling of disinfectant, and Daisy's flashlight lay by the head of her bed, as always. But today the galvanized tub we bathed in sat on the floor, along with a bottle of white vinegar and a contraption that looked like a hot-water bottle with a long, thin hose.

After a time, Daisy came down the stairs lugging a steaming bucket of hot water. She poured it into the tub without so much as glancing at me. Then she untucked the sheet stained with rust where it had touched the nails bursting out of the bubbled cement of the foundation and laid a towel in its place. She unbuttoned her pajamas and let them drop to the floor until she stood completely nude before me, skin hairless as leather and smooth as poured chocolate. I could feel my scalp prickle with fear and revulsion.

"Soon you gon' start to stink," she said, picking up the bucket. "I'm fixin for you to know what to do. You use half a jar of vinegar"—she demonstrated, filling the rubber bottle— "and a couple gallons of water." She lay on her back on the bed and opened her legs.

This is not real, I said to myself. I began to float, way up safe into the moldy wooden beams where the light leaked in behind the little curtained window. But I could not tear my eyes away from the shock of her opening. It was dark, nearly black, but the gateway gleamed a startling red. I sat, a frozen thing on my cot, as Daisy picked up the rubber bag and inserted the long hose.

She pushed it deep inside her and squeezed the bag, and water gushed into her, then back out into the tub. I closed my eyes and pressed my face against the cement, inhaling its ashen taste, cutting my cheek, anything to avoid the sight of my aunt's vulva. I imagined myself a bird that could fly through wood, through the hasty seams my father had made in the concrete, out above the quaking aspens to the gentle horses across the road.

At last the gush of water ceased. Daisy got up and began to towel herself off.

"All right, now you know. Get dressed. It's time to go clean the church."

~ ~

Today was Winter Carnival, a Steamboat tradition that had been started to inject a bit of fun into the year at just the point when it seemed winter would never end. There would be dog races and ski jumping and all kinds of spectacles after dark. The whole town would be there. Marjorie Perry, on her eight-foot skis. Local legends Loris and Buddy and Skeeter Werner, Olympic skiers all, and ex-Olympian Gordy Wren, who ran the ski school. Newcomers with Ph.D.s, struggling along on minimum wage. Even ranchers who rarely left their spreads would turn out, bundled up and cheery.

All week long the snow had been coming down dry and fine as swan's down, leaving a bowl of fresh powder in the Yampa Valley. The festival was going to be pretty as a wedding cake. My friends from Steamboat High would be up early, waxing their skis in preparation for a day on the slopes. At night they would all pile on the hay wagon, tucking themselves under scratchy wool blankets and snuggling together like a pile of warm puppies, while the Percherons with bunched haunches pulled them up the snow-drifted road, their breath hot on the air.

Daisy had always refused to let me go to the carnival. One year it was because it was too dangerous. Another year she said it was too expensive. But most of my classmates came from families of modest means for whom skiing was mainly a way to get around. The ranch kids did their share of the nonstop grind of milking, mowing, canning, weeding, worming, fence-mending, hunting, fishing, and butchering. They were lucky to go into town for a Saturday night square dance once a month. But even they got to go to Winter Carnival.

Maybe this time, if I was really good, Daisy would relent. I'd been so hopeful that I'd told my friend Lynne I'd meet her at the carnival. I didn't want to think about all the times I hadn't shown for things she'd asked me to. How many more times was she going to invite me?

The sight of Daisy hunched over the wheel like she was blind, driving down the road at fifteen miles an hour, set my teeth on edge. Yet I didn't feel right running her down behind her back. Complaining about her was like handing those she was convinced were arrayed "agin us" the very ammunition they were seeking. Nor was she as oblivious to what I felt about her as I'd thought. Just this week she had glanced over at me ducking down in the seat of the Jeep so my friends wouldn't see me and said, "You're ashamed of me, aren't you?"

I felt stricken. It had never occurred to me that she might notice. "Of course not," I said.

"Yes, you are," she'd replied. "But that's all right. You'll get what's comin to you one day."

Now I stared out the window at a mound of hay sitting in the middle of the field like a frosted loaf of sweet bread on a white plate. The weather was perfect, neither so cold as to be bitter nor warm enough to melt the snow to slush. The moon was going to be big, a polished steel smile giving the snow the sheen of satin. But we'd clean the church and go home, sit in the dark and say the rosary.

Something unruly and obstinate had taken root in me this year, like a calf who wouldn't go in the chute. I knew Daisy had made tremendous sacrifices for me, much as she exaggerated. But I wanted to be with my friends, whether that made me a selfish little wretch or not. Skiing was becoming bigger and better every year, and I wanted to be part of it. And Winter Carnival had been founded to cure just the kind of cabin fever that was turning me septic. The carnival was about making winter fun. But Daisy was suspicious of anything that didn't involve drudgery.

The sight of the church steeple lancing the blue sky, all that piercing fierceness, made my gut cramp. This was where we came on Sunday mornings, starving—nothing to eat before Communion. Daisy always took a pew in the very back of the church, jumpy about who would sit by her and who, like normal people, moved to the front or the middle. She secreted us at the very back in the movie theater too, against the farthest wall.

"See that? I been here thirty years and they still won't sit by a Negro."

Against my better judgment, I'd pipe up. "What if they just want to sit closer to the front where they can see better?"

"You shut your smart mouth. Don't nobody need to hear from you."

It seemed to me Daisy had gotten worse since Slim McCormack had given us a television set. RCA VICTOR read the outside of a box so enormous he could barely get it through the front door. "Don't get your hopes up," he said. "No tellin whether the signal will come over the mountain." He gestured with his thumb toward Mount Werner. He and Daisy maneuvered the box into the only bit of clear floor space, in front of my piano, then he looked around the room. "You don't got but two outlets?"

"Three," Daisy said. "But the other one is in the back."

"How you set for fuses?" he asked. "Don't know if this thing'll blow your circuits with that freezer on."

"What if we unscrew the lightbulb while it's on?"

"Not enough. You might have to unplug the Deepfreeze while you watch your shows."

I never would have believed Daisy would go for that, but without a word she walked over and unplugged the freezer, risking all our winter meat. In the sudden silence, our excitement pulsed. Slim threaded an extension cord across the back of the room and into the socket. When he turned the knob, the television hummed and clicked as its cells and tubes began sucking current. A grainy picture came to life, and a slushy sound emerged. Slim squatted on his heels, one hand in back of the set, turning knobs, while the other fiddled with antenna strands. Then he stood up and crossed one arm across his chest, propping a hand under his chin to think. He looked around.

"What you need?" Daisy asked.

"That," he said, pointing to a shoebox of used tinfoil. He wound the antenna in the extra aluminum and an ad for Kraft Miracle Whip blazed in sharply, making us all jump. Then I could feel my entire body drop into a state of hypnosis as I crawled into the tube with the lady in a shirtwaist dress, spreading the smooth white paste onto a smooth white piece of bread for her clean white child.

After that, Daisy and I fell into the television each night. Our favorites were the Western series like *Gunsmoke* and *Rawhide,* which ended happily every sixty minutes and reassured me that the cowboy values I treasured were revered across the nation. I saw my father in the heroes on the screen, the kind of guy who could handle the herd and a lady as well. Never mind that they were all white. Even the "Injuns" were obviously phony, with their heavy makeup and thick, stilted lines. But I was troubled by Daisy's new habit of scurrying home each night for the news. It was never good, and it always put her in a fury. She hardly breathed, watching Freedom Fighters with their arms linked, black and white, facing hoses, dogs, spit, and hatred. Television imported into our one little room the pestilence Daisy thought she'd left back in the South, and it accumulated there, like a low cloud that just wouldn't clear up.

I didn't know what a civil right was exactly, and I couldn't understand why blacks down there didn't just leave, like we had done. But I could see that the TV was making Daisy worse, wearing me out with her hypervigilance that read too much into a neighbor's courtesy wave on a country road, interpreting it as a snide dismissal or acceptance at last, depending on her mood. I developed an aversion to the news, and especially news of "the race problem," but I did join Daisy in watching the day that Lyndon Johnson—"and a damned Texan at that!"—signed the Civil Rights Act into law. We couldn't get the picture to stop rolling, but the sound was fine. Daisy didn't even cook the duck she'd thawed. Instead, she took out Robert Ball Anderson's memoir and stood in front of the television, swaying, tears dropping onto her shirt.

"I could go in any place in this country now and order a piece of pie," she choked. "I come to town to buy a sack of meal. White woman come out the store at the same time I was goin in and they pushed me off the walk into the mud." I marveled at how many years that scene had lived in her mind. It was as if she could not be available for anything else until that had been set right. "Caint none of them sons of bitches push us off the sidewalk now."

I wanted to touch her, to hug her, but she had never liked physical contact, and I didn't want to break the mood.

"They slaughtered us, Rita. Just like you'd go out and butcher a steer. Worse. They loved to harm you in your privates. White folks is a funny thing. They get like a pack of dogs that won't stop until they see guts spill. Now this man Johnson done put a stop to it. I never thought I'd live to see the day." She paused and looked at me. "White folks don't know it, but they crazy to the bone. Don't none of 'em make a lick of sense. Sometime, they ain't even human." Finally she walked over to the red chair and sat down. Then she looked up at me as though she were returning to the room.

"Who in the Sam Hill come up with the idea to set up at night in the colored folks' yard dressed in a bedsheet and set a match to Jesus' crucifix?"

I couldn't help it, I burst out laughing. And for once, she did too.

When we got out of the Jeep at the church, we could hear the band practicing in the distance for the parade that would open the night's festivities. Would my friend Trudy be playing her French horn in the marching band? I wondered how she kept her lips from freezing to the thing. Then I heard them testing the speakers, the words too distorted to understand.

Inside the church, Daisy bustled into the supply closet. She had just begun to run water into the mop bucket when Father Funk appeared, as if he had been lying in wait. Even though he had eyes, a mouth, a nose, I knew that if I saw him when he wasn't wearing his vestments I would be hard pressed to recognize him. During his sermons, my mind always scuttled off to gather wool on some livelier slope.

"Good afternoon, Daisy," he said brightly. "Oh, it's not

nearly afternoon, is it?" he corrected himself. "Beg pardon. I was wondering—well, it isn't only me . . ."

Daisy put the sour-smelling mop in the bucket and took her good sweet time before she faced him. She hung her head to the side, like the Carpenter himself, and regarded Father Funk with her wide, sad, blue-rimmed eyes.

"Of course, Father, anything."

"Would you mind—"

"Anything at all," said Daisy, with her most servile smile. The longer she smiled like that, the harder she would smack me later.

"Well, Daisy, it's the Stations. They, you know, need attention. What with the summer having been so dry and dusty. . . ."

Daisy's smile changed, though the same muscles stayed contracted.

"Of course, Father," she said. "That's Reeter Ann's job, but you know she caint see too good what with having only one good eye and me not having no money for glasses. It just costs so much and now she gotta have the Kotex. And a belt for the pad. And shoes. And then the tax. I just can't do everything. I gotta be her mama and her daddy both and don't nobody give me a dime. But I give her everything. Her own mama and daddy couldn't have done no better."

The priest began to edge backward as if he were a dog up against a badger. "Well, I know you have your cross to bear." I disappeared behind the sacristy, shivering with humiliation and anger. Daisy had never fucking mentioned that I should dust the fucking Stations. I kicked a cabinet, stubbing my big toe. Even though I couldn't make out what she was saying, I could hear Daisy going on and on. I knew the litany by heart anyway—a piano and piano lessons, not to mention three hots and a cot, and Daisy on her knees night and day, mopping floors, "and all so she can make something of herself."

I looked up at the pale porcelain Virgin with her foot squishing the naughty snake on the globe of the world. Was that the mortal coil you were supposed to long to shed? When I beat myself with a soft fist on the *Mea culpa mea culpa mea maxima culpa*, it felt like my heart tore. *Through my fault, through my fault, through my most grievous fault*. Mary with the white face stood on the globe of the world holding the Savior who died because of the dark daughters of Eve, who wanted that most terrible of all things—knowledge. And even though Daisy told me twenty-five times a day, "Stop asking them crazy questions," I burned with a desire to sort out this conundrum.

I was beginning to notice the men around me in an entirely different way: the texture of Slim McCormack's hands as he rolled a cigarette, cradling the Zigzag, judiciously rationing tobacco from the can, the petal-pink tongue flicking delicately along the pale paper, the deft stroke—up and back—that sealed the moistened edge. The scrape of the match-head with a cracked thumbnail. The tilt of the head away from the flame. The suck of the fire until the tip of the cigarette glowed orange. And then the plume of smoke exhaled, in a sigh almost of satisfaction. And the columbine-calm eyes that looked up and noticed me noticing him.

Other kids let off steam by going out for football or playing in the band or joining the cheerleading squad. I heard about the assignations that took place after school, long after I had departed on the bus or slunk down in Daisy's Jeep. But it wasn't only living far from school or Daisy's strictness that prevented me from joining in. I'd noticed the furtive kisses boys sneaked on the bus to speech meets, but I'd also seen how girls who kissed them back were judged by a different set of rules. And it seemed to me that a double shadow had fallen over me because I was not only female but black, and a current was humming through my body all the time, and there was no one to talk to about it.

Six months earlier Daisy had taken me to the doctor, worrying, "She's flat as a board. I'm afraid she ain't never gon' develop." But now that my breasts had begun to bud, she had grown wary and strangely quiet. Men and boys gazed at me now, even old ones, searing the back of my neck and the base of my spine with their eyes. It seemed to me there was something that everybody knew but nobody was talking about except in hot, urgent whispers I couldn't quite make out.

My mind wandered back to that boy at the Whiteman School, a private high school where Daisy and I had stopped to drop off some butchered geese. In the dining hall, I was amazed at the number of cute guys splayed over the tables in all directions. The girls had plush ski sweaters and expensive haircuts. They looked bored and mean, pulling on pink spun ropes of bubble gum and doodling. A surge of envy swept through me.

Then a flurry of energy rippled along the back of my neck. When I turned around, a tall blond boy at a table by the wall was gazing at me with his mouth open. It was as if his gaze entered my body and rode the circuit of my nervous system to the tips of my toes. I dropped my eyes, flustered and confused, but I could feel him continuing to stare at me boldly. I flushed. If I glanced back, would that make me a black whore? Just as Daisy and I were about to pass into the kitchen, I turned around. *Caught ya,* he mouthed.

I looked at the Virgin Mary again. If God made everything, then he made her between-the-legs too. So why the "immaculate" conception? Why wouldn't he, couldn't he, touch it when it was such a thunderously delicious thing to touch? What were the hydraulics of chastity? Who benefited from this harnessing of our power?

What about the day when I went up to Slim's cabin to return his wire stretcher? His truck was gone, but I called his name just to be sure. Then I tried the door, and it opened. I was only planning to drop off the stretcher and go back down and finish my chores early for once, instead of waiting till the fear of Daisy was on me good. But instead I stood a long time in the doorway, taking in the room, the scent, this man's way of being in his place. I pictured how he could pull barbwire with a stretcher in one hand, pluck a staple out of his mouth with the other, and hammer the strand tight in half the time it took Daisy and me working together.

A well-oiled saddle lay on the floor by the rough pine shelves, bridle and reins wrapped around the saddle horn. Motor oil and a funnel lay on newspaper by the kitchen table, along with a box of bullets—I knew they were .22 gauge because we used the same kind for the .22 of Gampa's I'd gotten when he died. On the back of the one round-backed chair, a sweat-stained Navajo saddle blanket aired—the scratchy kind, shades of gray and cranberry, with a flaw to honor the Great Spirit tucked into the bottom diamond, one red thread off center. Saddle soap sat by a pile of bills on the table, along with a coffee cup, a sugar bowl, and a square tin of Bag Balm. A beat-up coffeepot stood on the stove, burned black, two lonely cans of pork and beans and Vienna sausages on the shelf.

The crunch of gravel startled me, and before I could think, I bolted inside and closed the door, heart pounding so loud I could actually hear it. Sweat misted my brow and the back of my neck. Through the window I could see a black pickup go on up the road—one of the Swigert kids, probably heading up to the barn. Still I did not leave. Before I could talk myself out of it, I was darted down the hallway to Slim's bedroom.

Daisy liked picture windows, and there was one at the end of the room, a sheet nailed haphazardly above it to make a curtain. Plain unpainted Sheetrock barely covered the studs. Grimy sheets and Hudson's Bay blankets were turned back on the brown metal bent-frame bed, revealing the hollow Slim's sleeping body had left. The other side of the bed was almost pristine, as if he never so much as turned over during the night. A carton of Pall Malls, one end torn open, sat on the wooden crate he used for a nightstand, a half-full pack and a pile of wooden kitchen matches and a pint of Jack Daniel's alongside. I had no idea he ever smoked ready-mades. I pulled one out and smelled it—kind of sweet. I opened the whiskey and sniffed that too. I took a sip. It tasted like turpentine and I coughed, then felt the sudden peculiar heat oiling its way down my throat.

I lit the cigarette and lay down on the mattress. I practiced holding the cigarette the way I'd seen people do in the movies, my hand flared and graceful as a bird's wing, the smoke curling up, etching pictures in the air above the bed. Slim had driven big heavy nails into the wall to hang stuff on—halters, bridles, a couple pair of jeans, a plaid flannel shirt, a calendar from the hardware store.

The sheets were dirt-soft and smelled like sawdust, nothing like our sheets, which were stiff and sun-smelling after hanging on the line. I tried to blow smoke rings until I started to feel woozy, then I put out the cigarette in the coffee cup Slim used for an ashtray. I pictured how he stood patiently and let Daisy talk, without a glimmer of the agitation that tortured me when she made the same old infuriatingly vague references to her suffering—"I done had tragedies" or "I seen a thing or two."

I imagined the glint of light on his day-old scruff of beard, the smell of woodsmoke lingering in his wool shirt, the inward curve of a boot at the heel, the sling of his jeans on the jut of his pelvis, a thumb hooked nonchalantly in the pocket. In my mind he looked up and saw me watching, and our gazes locked. I felt impaled, entered, revealed. I could not breathe. I sat up, recovering myself from the trembling that swept through me in waves, the blood rushing hot and red in my ears. What was happening to me? I peeled myself off the bed and made my way back up the hall, out the door, and down the hill.

~~~

Daisy dragged the mop bucket into the sanctuary and tapped her fingers against the topmost edge of the font without actually wetting her fingers. I knew she was worried by the thought of hundreds of fingers in the same water, touching it to heads, breasts, shoulders, and finally mouths. She shuddered over toilet seats, doorknobs, glasses. In the South she'd seen whole families carried off by whooping cough and TB and measles and smallpox.

"Today it's gon' be your job with your iddy-biddy fingers to get in every crevice of the Stations and get 'em clean."

I couldn't find the words to say why this felt so hugely wrong to me. It seemed a sacrilege to touch sacred objects in that way. How could you still find mystery in an object you'd sprayed down with Pledge?

The holy water was cold. What made it holy? Was it still holy after the basin was scrubbed?

I gathered the rags and the lemon oil while Daisy began turning up the pews in preparation for sweeping and mopping. Cleaning the church was like seeing inside the body cavity of a dead deer. The smell of burned wax, the little red glass candles people lit to remember the dead—were they all merely things?

The first Station: Jesus Is Condemned to Death. I climbed a ladder nearly to the window and reached out my hand to touch the chipped nose of Pontius Pilate. The rag was too thick to fit into the crevice of the bowl where he washed his hands. Jesus was on his knees—"po' fella," Daisy would have said. I didn't want to draw the cloth across his rear end, but the dust, thick and gritty, had gathered there. *My Jesus, often have I signed Thy death warrant by my sins; save me by Thy death from that eternal death which I have so often deserved* went the prayer we said at the first Station. I'd never told anyone I often thought about killing myself. It was my little secret—I could get out. I only had to find the courage.

The next Station: Jesus Takes up the Cross. The cross at least was easier to clean, with its long planes. I could feel the weight of it. I forced myself to concentrate on the grit in the little nooks of Jesus' crown that had silted and hardened with age—it needed a bath in hot soapy water and a scouring with a rough brush. There it was again—I had to do something that couldn't be done and I would be blamed no matter what. I thought of the wine in the sacristy—just a thimbleful would help, like when I went down in the cellar at home and got into the dandelion wine in the heavy stone crock. I tried to scrape the dried dirt out with my fingernail and got a splinter instead. Blood bloomed and I gasped, struggling not to cry out, to remember that Jesus' pain was the point, not mine. I leaned against the pew, cupping my hand, inadvertently turning my back on the crucifix. It was as if my fear of touching these things had been validated.

Daisy muttered to herself as she wrung out the stringy gray mop. Was she praying? The windows let in soft, pale light. I dragged my stool along the wall to the next Station.

A god, if there really was such a being, could simply make all things well. But if Jesus Christ could not escape this kind of torture, what lay in store for me? It seemed that all roads led to blood and shame, sorrow and death, so what was the point?

The radiator clicked on. That meant it would soon be time for four o'clock confession, and I was only at Station Ten: Jesus Is Stripped. The women stood at the foot of the cross, heads bowed, figures draped. I noticed they had left His crown on. I felt as if I were suffocating in sadness, drowning in it. Five more to go, and the light was already dropping. Right now, I bet those Whiteman boys and girls were piling into the van back to the school, cheeks red from that final run down the slalom.

I took the dirty rags and cleaning supplies back to the closet, washed my hands, and went back inside the church. Time to go to confession. I reeked of lemon oil, dust, and a week's worth of sweat. On a normal Saturday night, Daisy and I would pour buckets of water into the galvanized tub and bathe, shampooing our hair and pressing it so we'd be clean and sleek for Mass on Sunday morning. But this wasn't an ordinary Saturday.

I heard the door at the back of the church open and prayed it wasn't Daisy. I wasn't ready to see her slump in for her devotional. This prayer at least was answered. Just from the soft rustle, I knew it was the lady who came every day at four on the dot. Without even turning around I could picture her mauve coat and the head scarf tied under her chin, I listened to her buttons clicking against the fountain and the rustle of her gloves as she pulled them off to bless herself with the holy water, then the squeak of her overshoes as she genuflected and came down front to light a candle for the bay of votives, flickering in glasses the color of dried blood. The long match rasped as she lit it and the flame spliced the gloom. I knew she must be thinking of her son, who had died when we were in sixth grade. Had she been angry with God for taking him? Was she still? She seemed so docile, as if all her energy had been drained out. I had never been able to bring myself to light a candle for Mama or Gama or even mean old Gampa.

In the distance, the microphone blared again. I knew the town kids would all be giddy now, tracking down lost mittens and earmuffs. I wondered if there was any chance at all I could beg hard enough for Daisy to let me go to the carnival, even though the priest had upbraided her about my cleaning. I had already picked out my clothes. A second-hand raccoon coat rubbed bare at the corners—but that wouldn't show at night—and a hat I'd knitted in home ec from the wool of a fine black ewe, a prizewinner at the Routt County Fair. Anybody who'd been there would recognize that yarn. Would Marie get to come into town? Lynne's nose and cheeks always got so pink in the cold, and her mom made the best hot chocolate. It just wasn't within Daisy's realm of possibilities that I should be friends, genuine friends, with these kids. The idea of me being outside her control was too unsettling.

The face of the Madonna looked gray in the flickering light of the candle. I couldn't bring myself to get down on my knees. *Blessed is the fruit of thy womb.* Nothing—no single thing I could ever do—would be worse than to have a baby, according to Daisy. If I became a nun, completely walked away from life, would I find peace?

I struggled to find some speck of contrition within. I couldn't afford much more if I wanted to remain me at all. What would I confess? Covetousness? I had certainly coveted my neighbor's skis, and her Breck hair, and her mother and father and indoor plumbing. I had coveted her horse and the proper tack with which to ride it and the barn she kept it in. I had coveted my neighbor's yellow flowered cookie jar on a clean kitchen counter with everything set just so.

Daisy came out of the confessional to find me still sitting in the pew thinking, neither kneeling nor preparing to go any-where. She nodded her head toward the curtain.

I pretended to be deep in prayer.

She genuflected and slid into the pew. Then she knelt beside me, bent over, and whispered, "If you going to confession, you best be about it."

"I'll go tomorrow," I said. Daisy looked at me open-mouthed with astonishment, but Father Funk emerged from his booth and it was too late. I felt a small surge at this pitifully small rebellion, dearly as I knew I would pay later.

~~~

On the drive home, I could see from the tightness in her jaw that Daisy had set herself against my going anywhere that night.

"Snow coming," she said. "It's just too warm, don't you reckon?"

I looked at her. Her bandanna had tilted off to one side and her braids were springing loose. I itched to pick a fight with her, to snap, "Just give it a rest with the Farmer's Almanac crap." But I held my tongue.

We passed a dog team headed for the carnival—a couple of German shepherds, a malamute, four huskies, and something that looked like a wolf. How I wanted to go with them!

And then it erupted out of me. "I can't believe you men-tioned my Kotex to the priest!"

Daisy gripped the wheel and made a point of ignoring me.

"And you didn't have to tell him I was half blind. He doesn't care about that. I really hate how you talk about me, you know? Like I was a thing and had no feelings at all."

Just as the words rolled out of my mouth and away, and with them my last chance of going with my friends, the hay wagon came jingle-belling around the corner as if to taunt me. Daisy pulled the Jeep to the side of the road to let the team of dark Percherons pass. Nearly seventeen hands high, they looked curried and festive in their shiny brass harness.

The driver pulled on the reins and brought the horses to a halt. "You all comin in tonight? It looks like we're going to have a good crowd."

Daisy put on her singsong voice. "No, I just don't see how we can do it. We got chores and so on and whatnot. And tonight, me and this child got to pray."

I got out of the Jeep, slamming the door. The sky was completely dark now, and cold. Daisy was wrong. No more snow tonight.

I knew they were both looking at me. With an effort, I stuffed myself back into a size that would fit in the Jeep and closed the door, not daring to glance at either of them. All I could do was rock and suck my splintered thumb.

The horses shifted, making the tack squeak. Daisy and the driver both spoke at once.

"Well—" Daisy started.

"Well," he said, "I guess we'd best be getting long—"

"You all have a nice time," Daisy said. "You sure got a pretty team there. Reminds me of when I was living with my husband in Nebraska."

If the driver didn't leave in a hurry, I knew she would start in with her stories of locusts and sod houses and living on jackrabbits and we would all be trapped here needing to pee until Christ came back. I plugged my ears with my mittens and hummed all the way home, because I couldn't trust myself to speak, or even listen.

When we got to the gate, at least three feet of hard snow

blocked our driveway. The snowplow operator had not lifted the blade when he went by.

"See that? Them rotten sons a bitches won't give a colored fella a break."

Daisy parked the Jeep by the side of the road and fumbled with her keys, trying to unlock the gate.

I screwed up all my courage. "Daisy, may I please go in and use the phone?" I was not allowed near the telephone. It wasn't just the expense. Daisy imagined the whole neighborhood waiting by the phone for an opportunity to listen in on our business. She had even devised a code to use with her brothers, so that no one would know what she was talking about. But it was worth one last, desperate try. "I need to call Lynne so I can tell her dad what time to pick me up."

"You ain't going no place. Don't need to use no telephone. Now get them snow shovels and let's get down to business."

I got the shovels and we set to it. I could picture Lynne waiting for me to come and then giving up with a shrug. I slammed the blade into the snowbank.

"You know, you getting to be one of those hotheaded Negroes, just like your father."

I couldn't stop myself from rising to the bait. "Daisy, you promised. A whole month ago, you said I could go. Please. I promised Lynne I wouldn't say I was coming and not show up this time."

"I ain't saying another word. You can't go, and that's final. Now close your mouth."

We shoveled in silence. Jiggs, sensing the anger, quit barking, whined once, and lay down with his tail across his nose.

The blade sliced through the snow and I heaved it farther than necessary. I knew I should let it go, but I couldn't. "Daisy, why did you lie to me?"

Her eyes widened and the pale blue rims around her irises seemed to get lighter. She was panting. Finally I'd gotten her as angry as I was.

"Nigga, did you just call me a liar?"

"Know what, Daisy?" I replied. "I'm not a nigger, and you did lie."

She gripped the shovel handle and raised herself to her full height. But now, I realized, I was as tall as she was.

"I won't hold still for no sassin round here. You a nigga if I say you are."

"I am not, but you are. Look at how you dress. That stupid red bandanna. How you crawl in front of whites. How you beg for food. How you give everybody white the best fruits and vegetables and leave the stunts for us."

She put up her hand to smack me, but I danced back out of range. We both realized I was too big to hit anyway. Daisy's face crumpled like cottage cheese turning to curd. "After all I done for you. Give you a place to sleep. Mop floors so you could get a good education."

"You lied, you lied, you lied!" I threw the shovel down the road so I wouldn't hit her with it.

"Rita Ann Williams," Daisy said. "It's time for you to go."

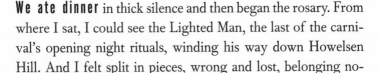

We ate dinner in thick silence and then began the rosary. From where I sat, I could see the Lighted Man, the last of the carnival's opening night rituals, winding his way down Howelsen Hill. And I felt split in pieces, wrong and lost, belonging nowhere, with nothing to believe.

7 ~ Mount Saint Scholastica

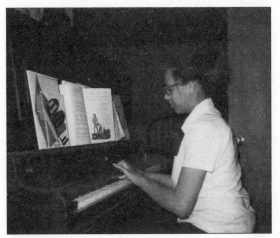

At the piano

~~~~ **After that scene** with the shovels, Daisy and I
were both sorry for our own reasons. I was terrified at the possibility that she was going to call the sheriff and send me to a reformatory. Daisy was devastated to discover I was just another
"hotheaded nigger" bent on dashing her dream of making me
into "somebody." I had dashed my own less-than-noble dream
too—that of besting my errant sisters.

Monday morning Daisy woke with silent determination, boiled up a pot of lumpy oatmeal, and served herself a full bowl without so much as looking at me. Then she pulled a kitchen chair over to the little table where she kept the phone.

"Father Funk, I've got to git rid of Reeter Ann," I heard her say. "Can you tell me some places to send her?" Shame and fear boiled up inside me. "Yes. Yes. Well, she *used* to be so sweet. Now she's got a mind of her own. I won't stand for it." As I listened to the scratch of her pencil on the tablet, a tiny voice inside me pondered this conundrum: How could I get an education and learn to think for myself, as she was always admonishing me, if I couldn't have a mind of my own?

Finally, she got through to some place in Canon City. I recognized it as the town where Ernest had done time for rustling cattle, or pigs. Daisy had said he'd come out of the pen with "a roll of bills thick enough to choke a horse." And Canon City was near Pueblo, where they penned up the "mentals." Icy terror numbed my body and I sat on my bed motionless, hardly breathing.

"No, Sister, I can't wait till the fall. She's got to go now."

Outside, Jiggs's chain bumped against the porch. My vision narrowed, and I stopped straining to hear. I knew enough—Daisy was about to abandon me on the doorstep of Holy Mother Church. "All right then. We'll be catching a bus directly." The phone receiver settled back into its cradle.

After that, it was all about practicality—what to pack for that warmer part of Colorado, two hundred miles away. By the time we got on the bus, the silence stood between us like a lead wall. Daisy looked out the window as we passed the stands of pines that clothed the pass. At last she spoke. "You got a lot to buck. I hope you can at least get a good education." I longed to tell her I was sorry. I wished she'd hold me, just once, and when she dozed off, I leaned against her arm. How odd it was that such a tough old bird had such soft, warm skin.

Canon City didn't look like much. It was about the same size as Steamboat but dowdier. Frumpy buildings squatted on a flat plain with the Colorado State Penitentiary slammed up against the butte above. We caught a cab to the academy at the edge of town. I had expected something medieval, with spires and a tall iron wall, but the building absolutely sparkled with white marble. Inside, the foyer sported a quantity of white stone and the furniture gleamed with fragrant lemon oil. A stylish, modern cross hung above all, and there were circular wax marks on the floor, as if the polish itself had been polished.

While we waited in the lobby, we marveled at the sumptu-ous couches, the bright, spacious feel of the room. Daisy picked up a brochure that listed the things a Mount Saint Scholastica girl would need. "Uniform?" She shook her head. "Lordy, what else? And three or four blouses. I can't afford but one. You'll just have to wash it out in the sink at night and iron it in the morning."

A flock of nuns came swooping past the open door, habits flying, beads clicking, followed by a fan of about twelve clear-faced girls in immaculate white blouses and pleated skirts. The spirit of these sisters seemed cheerful and energetic, completely different from those I had met in Steamboat. One of them was quite pretty, with big dark eyebrows and huge blue eyes. The entire group, nuns and girls, looked healthy, well fed, and safe, unacquainted with perversion and grief, despair and martyr-dom. Daisy and I stared at them as though they were another species—not just white people but Blessedly White.

One redhead—the very picture of Ireland, with freckles on a cream skin—stopped long enough to pop open her notebook. A fancy three-hole plastic case with lots of pencils, a sharpener, and a brand-new eraser. Envy pierced me. I snuck a glance at Daisy and caught a look of longing on her face that paralleled what I was feeling.

"Ain't that child just as perty as a speckled pup?" Daisy said, openly expressing the awe I was too proud to show.

The Irish girl and a tall, Spanish-looking girl glanced at us and put their heads together, giggling. Fear and anger flushed through me. Were they laughing at us? I looked at Daisy, sitting awkwardly on the edge of the couch as though she was afraid she'd dirty it. Her body showed the signs of having been worked past reason or health—the bunioned feet in bad shoes, the knees broken down by bucking bales, the nails perennially rimmed in black, no matter how much she scrubbed them. She touched everything in a gingerly way, as if coming into "the white folks' house" uninvited. At the front of her jeans the hernia bulged where she had literally busted a gut, and her breasts sagged in her secondhand brassiere. At least her jeans, like mine, were clean and as wrinkle-free as the long bus ride allowed.

I jerked my chin up and hid my own calloused hands, pulling myself up tall enough for the both of us. "Daisy, quit staring. Jesus Christ!"

Daisy looked alarmed. "What is wrong with you? Got your mouth all pushed out like a nigga. Straighten up and act right for once."

At that moment, a great huge nun bounded in. She had big pillow bosoms and sausage fingers but surprisingly tiny feet, her shoes as shiny as a beetle's back. A tightly pleated coif seemed to squeeze her features into the center of her face.

"Welcome," she said. "I'm Sister Mary Bernard, the mother superior of Mount Saint Scholastica. Come into my office." I allowed myself a small measure of hope. But the level of fear that

I was suffused with made it hard for me to listen as she and Daisy worked out the details of my stay—I would earn part of my keep washing dishes, Daisy would send monthly payments. And then they both rose and looked at me. It was the first time Daisy's eyes had met mine in a week, and they seemed full of defeat, and relief.

And then she was gone, and Sister Mary Bernard led me down the hallway, a hundred feet of shining white tile and ivory walls adorned with a tastefully small crucifix. As we passed the dining room and then the library, she fired a long list of details at me, unaware that I was so overwhelmed by the sheer luxury of the place that I could hardly hear a thing she said, the words bouncing off me like blows off a cartoon character. She led me into a room the color of boiled bone, and I could feel my whole body quiver in response to the implicit luxury of light switches, thermostats, drinking fountains, bathrooms.

Sister opened a dresser, a closet with hangers, a built-in blond wood dresser with three drawers. "Your things here. Your books there." I was to have my own little desk! I moved to it, touched the smooth wood.

The sister's face was that peculiar mottled color of white people who shun sunlight, not unlike the breast of a plucked goose. I could see she had to work to support the weight of the fabric falling from her headgear. She was making pleasant noises at me, and I knew I should focus on what she was saying, but I couldn't hear very well for the blood rushing in my ears. Finally, she stopped talking and just looked at me.

"Yes, ma'am," I said, realizing that she was waiting for me to say something.

She smiled. "We say Sister here. No need for 'ma'am.' So I'll see you downstairs, then?" She reached out an immaculate hand and patted my arm, then swirled away in a small tornado of black yard goods.

I looked around the room, then sat on the bed and tested the mattress. The room was far too white to sleep in. I opened the drawers, where a faint smell of perfume lingered. No fly specks on these windows—I bet they didn't dare.

~~~

Two of my roommates turned out to be from Mexico. The other was a gangly, friendly blond from Denver named Elizabeth. All the other girls I saw in chapel the next morning were the color of fuzzy new peaches. Under the soaring ceilings, the stained panels of glass like jewelry, everyone knelt in rows—the sisters first, more than a dozen of them, and the girls in uniform behind. Here the sisters hardly appeared human, more like statues with folded hands, downcast eyes.

Afterward the girls erupted down the hall to the dining room, suddenly full of giggles and bounces. The smells that emanated from the room were nothing compared to the sight that awaited me: tables full of bone-colored china and gleaming silver, bowls of steaming scrambled eggs, fresh bread and cinnamon rolls, pitchers of orange juice beaded with condensation, and as much fresh milk as anyone could want. And it wasn't even Sunday. Saliva gathered at the back of my tongue while we waited for a sister to say grace. When she finished, she fastened her direct, steady gaze on me. "Everyone, say hello to Rita."

Elizabeth said, "That's Sister Mary Richard. She's the coolest." Then everyone sat down, the other girls all talking at once. Several sisters trundled in with bowls of hot cereal. I could hardly believe how they feasted, especially when there seemed to be no heavy work at all.

The creamy redhead I had seen in the lobby the day before introduced herself as Caitlin from Meeker. "I'm from Steamboat," I told her. "Really," she said, her forest green eyes animated. "I've done barrel racing there." She went off on a horse tangent involving an Appaloosa mare named Moon. And as if I'd been taken over by a little alien during the night, I countered with, "My father works on a ranch in Wyoming" and went on to describe the palominos with the long blond manes and lush tails, the hay and the sun, savoring Caitlin's envy.

"Are all nuns this rich?" I asked, changing the topic.

"For criminy's sake, Rita," she said. "They take a vow of poverty. Poverty, chastity, and obedience."

But they didn't seem all that poor to me, and we were the obedient ones. Why was I the only one who ever seemed to question these things?

After a few wide-eyed responses, I learned not to talk much about where I came from. What would it gain me to reveal that my aunt had only just decided, in her mid-sixties, to give up carrying out slop jars and chopping ice down at the creek when it was thirty below, and install plumbing in our house? Besides, classes here were so much more intense than those in Steamboat that soon I had hardly any energy for anything but my studies. Classes began at eight and ended at five. Study hall started at seven and concluded at ten. I was already a semester behind in Latin and in algebra. The huge library felt like its own kind of chapel, and it was always open. I could go in any time of day or night, and open any volume I wanted.

I resumed piano lessons too, with Sister Mary Joseph, whom I had heard play with surprising vigor and authority at the opening assembly. She was tiny as a sparrow, all the bones of her hands visible through the pale white skin. It was amazing to see how rapidly her tiny fingers covered the keyboard. At our first lesson, she asked me to play something and I chose "Für Elise," which my fingers still remembered from my recital a couple of years before. But I was nowhere near the beat. It wasn't that Mrs. Gilbert hadn't taught me to count. It was that I played the easy notes all in a rush, counting all the while, and lurched more slowly through the hard ones, counting still. Maybe Daisy's stick of wood always at the ready had obliterated in me all hope of concentrating on anything other than getting hit.

Sister Mary Joseph sighed and turned on the metronome. "One dee and dee two dee and dee," she chanted, in rhythm with the beat. "Try it again."

I lurched through the piece a second time, with no greater success than before. Sister Mary Joseph stood and walked to a shelf by the window. "You've got to learn to play one note at a time. You're getting out ahead of yourself." She pulled out a book of sheet music. "Let's try some Bach." She opened to the first prelude. "I want you to work only on this one page this week. Play as slowly as you need to, but cleanly. With the metronome."

In the weeks that followed, I faced down my fear of that first page. I realized that the tension embedded in my fingers ran all the way up my arm into my neck and that managing even a bar of notes was impossible unless I learned to relax. Every time my concentration wandered, I lost count and began again. But week by week I settled down inside the music, learning to follow it as if it were a map—one step, one note. Bach's pieces were so mathematical, built so logically, in such interesting increments, that they gained momentum as they cantered along. What an odd realm of serenity lay inside the music now that I could read it, someplace I could travel to anywhere there was a piano. All I had to do was practice.

Soon I began to regard life at Mount Saint Scholastica as normal. The chores that had occupied our every spare moment there were magically and seamlessly taken care of here. Still, the luxury of padding to the potty in the middle of the night on a warm, clean floor, and coming back to a nice firm mattress, the predictable routine of food and heat and light and prayer in the pretty chapel—didn't exactly seem to be mine. It seemed to be happening to someone else, someone I was mimicking—someone who actually belonged.

I couldn't stop thinking about Daisy, though, and contrasting the world I now inhabited with the life she led. Like Daisy, the nuns rose early and labored all day, each in her shroud of humility. But while they had their specialized jobs within this tidy little community—this one a baker, that one a librarian—Daisy struggled frantically with a thousand details every day, all on her own. She cleaned eight different houses and hunted and fished and shoveled snow or dug weeds or mended fence or carried water, chopped kindling, made soap, washed clothes, guided them through the wringer, pinned them on the line, brought them in, canned and preserved fruit, skinned and gutted entire deer and elk, rabbits and antelope, butchering them up into packages to take to the locker in Steamboat. And then she was off to tend to geese, chickens, ducks, guinea hens, maintain the henhouse, weed and tend the vegetable plots. Not to mention dealing with a dog she didn't like and, until now, a child she didn't want.

There was another difference too. Daisy seethed with conflict about the tedium and exhaustion of her work, bitterly recalling the days with Mr. Anderson when she had been wealthy.

And much as she would have claimed otherwise, she considered herself a martyr, confident that when the sheep were to be separated from the goats up yonder, she would not be found wanting. If the Sisters of the Order of Saint Benedict were proud of their humility, they did not trumpet it.

I kept these ruminations to myself, however, because I was genuinely grateful to be out of the vortex of Daisy's chaos and in a routine where what was expected of me seemed doable. I liked eating good bread and bathing every day and never once worrying about how soon the Klan would come and crucify me on a flaming cross. On cold nights I reveled in the central heating that came up through a grate in the floor without any effort from anyone, standing over it until my ever-freezing feet got too hot. I secretly squandered electricity, deliberately not turning off my lamp when I left the room, delighting in the fact that I could return to its warm, welcoming glow an hour later.

The one thing I could never get over was the abundance of water that did not have to be fetched from the creek and heated on a stove that had to be fired up and tended to, the ashes toted out. I loved not having to carry my waste and sacks of lime to an outhouse in the snow. I loved settling into the bath every single night, letting my body submerge completely. Gradually, my nails got clean and stayed that way. And the calluses on my palms from handling shovels, ax handles, and pails began to soften and disappear.

Even the perturbing randiness that had begun to rise up in me in such a troubling fashion found safe harbor, fastening itself upon a true man of God who would not take advantage. Like every other girl at the place, I plunged headlong into love with Father Francis, our priest. He was Mexican, with skin the color of butterscotch and long straight eyelashes that nearly hid his big dark eyes. I wanted to put my tongue on the pulse at his throat, near the white collar, and lick up his neck until I found the mouth as dark as a cluster of black currants. On Saturday afternoons when I pressed my face against the grill of the confessional and smelled the scent of lemon and incense that seemed to hover about him, all my plans to make a good act of confession took flight and I could never remember a single sin.

Sometimes, Father Francis taught religion class. When he looked down at his notes, his hair fell forward like a curtain of black satin, and when he looked up and brushed it back, the entire class held its breath. When he lifted the host high above his head on Sunday morning and said, "God is love, and he who abides in love abides in God, and God in him," his words washed over and through us as truth.

I began to relax, cut up a little. Caitlin had snuck in a little transistor radio she liked to listen to before dinner. Sometimes she'd lend it to me after lights-out and I'd listen to rock 'n' roll, or tales from the outer world. For the first time I began to get interested in the news, not only the plight of black people but that of other people as well. I fantasized what it would be like to take a stand for what I believed in, instead of keeping my head down and my mouth shut, the way Daisy told me. I imagined myself going to the bus station and riding all the way to the steps of the White House, where I would douse myself in kerosene like those priests protesting the war in Vietnam. I would be a real martyr, at only thirteen. I would be canonized, a second Saint Rita. Then Daisy would find out and be sorry she had sent me away.

One spring evening Caitlin was visiting Elizabeth and me in our room while we waited for dinner, when a new song came on, about doing the Twist. I didn't know exactly what the Twist was, but without thinking, I began to dance, putting into play all those undulating moves I had seen at Perry Mansfield. I closed my eyes and gave myself over to the music.

When the song ended, I turned around and saw my friends standing rigidly at attention and looking past me. Instinctively, I reached over and snapped off the radio. I turned around, sensing before I saw her that it was the little old sister whose name I could never remember, her hands folded under her bib. Her mouth was open and she wasn't even breathing. She was angry, yes, but there was more. She was horrified.

"Exactly what do you think you are doing?"

I didn't know how to answer that. It seemed obvious. I had been dancing.

"Whose radio is that?"

Not wanting to get Caitlin in trouble, I said, "Mine."

"Let me have it." She snatched it and whirled out of the room.

What would happen now? Would I be kicked out? I nearly blacked out with apprehension. But even in the midst of my shame and fear, I sensed that it wasn't dancing that was the problem; it was being caught in a moment of pure and abandoned bodily pleasure.

The reprimand I feared never came, though, and before I knew it, the term had ended. Suddenly all the other girls' closets were open and their parents were filling the driveway two by two, like Noah's ark, and filing up the stairs to collect their fledglings. Caitlin's parents were among them. She had horsey parents who clearly had come willingly, and she went happily, and it was amazing to see their sincere concern for and interest in that daughter of theirs with the flame-colored hair.

By vespers, all the girls were gone, and the dorm took on an eerie emptiness. The wind toyed with a shower curtain, and the arc of light from the prison steadily scraped the darkening sky. I didn't want to ask why nobody had come for me.

Finally, Sister Mary Bernard called me into her office.

"Daisy has asked us to keep you until summer camp starts," she said.

"Am I going to summer camp?" I asked. I hadn't even known they had one.

"Indeed you are. Isn't that just wonderful?" said Sister Mary Bernard in her habitually adamant way.

I nodded dubiously, trying to comprehend the notion of summer camp at this proper school.

"It just didn't make sense for you to go all the way to Steamboat for three short weeks only to turn around and come right back."

"Are there going to be any other kids here until then?"

Sister shook her head.

Daisy not wanting me to come home stung. Even if I did not want her, I wanted her to want me. But Daisy was always a step ahead. I bet no one had ever made such a request of the nuns. I could hear the hard-luck story she must have spun, about the cost of the bus fare and the eight houses to clean and her failing knees.

Sister Mary Richard gave me the job of dusting books in the library. I continued with my dishwashing job, took my meals in the dining room with the nuns, and slept in the oddly empty dorm. The library I had entirely to myself, and plenty of time to explore it. I could use the practice rooms too, and sometimes even the piano in the auditorium, where I had first heard Sister Mary Joseph play.

It felt vaguely illegal to take a seat at the big, blanketed grand piano, although I couldn't say exactly why. I tried a C scale, and the first tentative notes sounded dead in the vast space. Then a three-octave F scale, with greater confidence, remembering to tuck my thumbs under in time so that the rhythm didn't lurch. Outside the water sprinkler clicked its spray across the lawn in long, drowsy sweeps.

The next day I brought my music, and pulled the blanket off the piano and raised the lid. This time the notes ricocheted off the wide walls, and still no one came.

One day Sister Mary Richard came upon me in the library, reading a geometry book. I hadn't studied geometry yet, but the triangles in the book reminded me of the braces Slim McCormack built to reinforce our fences. It intrigued me how he knew at which angle to nail them.

"I'll show you," she said. She took me to a classroom and drew a diagram of a flying buttress. "You don't know about physics yet either, but you will. Algebra, physics, trigonometry, and geometry all are tools for us to measure our world, how it works," she explained. For the very first time, I began to see the point of math. In Sister Mary Richard's hands, the numbers revealed their connection to the Bach I was learning, and to the rest of the world. We established a pattern of meeting in the classroom whenever her schedule allowed. She would speak to me not only of math but of the motherhouse in Chicago as well, and how the sisters left family life when they became nuns, dedicating their lives to service. What a relief that must be, I thought.

Sister Mary Richard was nothing like the nuns I had known in Steamboat. She seemed to be a full person. She wasn't particularly pretty, but she was commanding, with terrific intelligence and a sincere compassion. She made me wish I had a vocation.

With little to do, I wandered the acres of manicured lawns

and explored the outbuildings. Neither a student nor a novitiate, I was in a kind of limbo, my sense of time distorted by anxiety and ambiguity. I imagined myself dressed in black, in the order of Saint Francis, working in Africa on a reserve, treating elephants and lions. I thought how lovely it would be to lead a simple life, to concern oneself simply with being a deeply good person who cared for the earth and the creatures. I began to think about becoming a novitiate. The problem was, I wasn't sure I was worthy. I wasn't certain I believed.

One day I came upon a red brick building that turned out to be the laundry. It was hot as an oven inside, and a very short nun I recognized from chapel was operating a mangle that was almost as big as she was. It emitted bursts of steam as she clamped it down on a yard of black fabric.

I leaned against the doorjamb, fascinated. When she saw me, she jumped. And then she laughed.

"You ting. You scare me to det." Her speech had the music of German in it.

"*Wie geht's*," I said.

She broke into a smile of joy. "*Du kennst Deutsch?*"

"*Ja*," I said. "I speak German."

And then she rattled off a battery of speech that I couldn't understand at all. "*Mehr langsam*," I said. Slower.

"*Ich bin Schwester Cordelia. Kommst du hierher später und wir können eine Tasse Tee trinken.*" Come back later and we'll drink a cup of tea.

From then on, every evening at dusk, Sister Cordelia and I would sit on the steps of the laundry, watching the light soften the butte behind us and reading German poetry together. She had a terrible old tin teapot to which she kept adding bags until the tea looked like swamp water, but I drank it anyway. Her favorite poem was the one that possessed me from the moment I heard it, *Die Lorelei* by Heinrich Heine. The first lines always grabbed me:

Ich weiß nicht, was soll es bedeuten,	I don't know what should be the matter
Daß ich so traurig bin;	That I am become so sad;
Ein Märchen aus alten Zeiten,	There's a tale from ancient times
Das kommt mir nicht aus dem Sinn.	That I can't get out of my head.
Die Luft ist kühl, und es dunkelt,	The air is cool and it sparkles
Und ruhig fließt der Rhein;	and the Rhine flows peacefully by;
Der Gipfel des Berges funkelt	the top of the mountain is glittering
Im Abendsonnenschein.	in the evening sunshine.

She didn't know much English, but she remembered the melody that went with the poem, and when I had mastered the words, she taught me the music. Her thin, scratchy voice didn't bother me. She had actually seen the Rhine! I imagined it looked like the Yampa River back home, and that the mountains in the poem were similar to the Rockies.

One night I picked up a nun's white coif she'd been pleating with her permanently reddened fingers and held it above my head. "Can I put it on?"

All the merriment ended.

"*Nein, Liebchen. Dieses ist nicht für Ihnen.*" She took the coif from me.

Why wasn't it for me? "Don't you think I would make a good nun?"

She sighed. "Oh, Rita. Wenn I come here my family have no money. There is nothing for me to do. Where could I go with six sisters? But you. You could do things. Get an education. I can barely read. And children. *Mein Gott*, you'd have such pretty children. Get married and have a family. A full life." She stood up and turned her face to the wall, and I realized she was crying.

One afternoon the week before camp was due to start, Sister Mary Richard sought me out in the library and told me she needed to speak with me. I was vaguely alarmed. Usually when someone wanted to "speak with me," it was because I had done something wrong. But I had been particularly virtuous that week, in chapel on time, my bed already made.

I waited for her in our usual classroom, the summer sunshine bleaching the walls a blinding white. As always, Sister Mary Richard arrived in a rustle of skirts, as if she were being carried in on the wings of virtue itself. I felt a surge of happiness, but Sister's face was serious.

"Rita," she said. "I have something to tell you that you are not going to like very much."

Right away, I knew they were going to kick me out. I shrank back and started twiddling my thumbs. I counted the tiles on the floor. Four this way. Eight to the door.

"I thought this would be easier on you coming from me. You will participate in everything at summer camp next week with the exception of swimming. As you know, we don't have a pool here at the academy, and the owner of the motel where the pool is says he can't allow any Negro children to swim there."

The word *Negro* ricocheted around the room like a bullet. My mind clutched for safety in the mundane details about me. The light that bored its way through the thinly threaded curtain. The shadow of her habit as it spilled onto the floor.

When I said nothing, she seemed to grow anxious. "I am truly sorry. I can see you are taking this hard."

What could I say? I was trying to remember the moment before she had spoken, trying to retrieve that time when it still made sense to do my very best. I felt sorry for Sister Mary Richard. She had been nothing but kind to me, but in the end all her efforts were going to be useless, because everywhere I went, I would always be a problem. Race would be always be a problem, my problem, even in this place of God.

Yet even in the midst of my concern, something else was rising.

"Sister Mary Richard," I asked, "what made you ask this guy whether I could swim there? Why didn't you just take me?"

She looked miserable. "Well, Rita, I just didn't want to get you there and have all the kids around and have him ask you— ask us all—to leave. Wouldn't that be worse?"

"I don't know," I said. I still wondered what had made it be a question to begin with. "What made you wonder if I would be good enough?" I asked.

She flushed. "It's not that. It's definitely not that." Her face was the very picture of consternation. "Oh, honey, you're hurt, aren't you?"

Something in me snapped then. I walked out on Sister Mary Richard without another word, something I had never done to anyone, ever. All I could think of was finding a place where I could be alone, where she wouldn't think to look for me.

At home, when Daisy was on a grind, I used to go out into the forest at night and sing myself sane. If the idea of a cathedral worked for me at all, it was in the presence of those aspen stands, even more so among the spruce. I was homesick for the messiness of the forest floor, the coolness of petals, the rustle of wings, tiny paws on rough bark, icy water smoothing rock, the sense of collective breath held. I felt something holy in the outdoors back home that I had never felt in church.

I ran behind the auditorium as if this thing pursuing me could be outrun. The crew-cut grass was organized between sidewalks, which seemed to me an odd waste of land. No vegetables or scented flowers, and certainly not the kind of verdant pillow that could feed dreams. I never saw anyone sit on it or lie back on it to watch the clouds. It didn't even smell good. I had been shocked on my arrival at how many dandelions had taken hold, and surprised to learn that no one plucked them out. I guessed they had no idea that lopping off their heads with a lawn mower once a week only fed the strength of the root. In fact, I imagine I was the only one who noticed them at all.

It was beginning to seem that much of my survival among white people depended on not noticing things. Not heeding when the fallopian tubes sent out eggs and hormones engaged, calling for seed. Not hearing the drumbeat. Not making a peep when Daisy handed me over to the nuns like a goose she'd gutted and plucked.

I heard the motor of the tractor-style lawn mower, ready for its next sweep. I could see that the driver was black, wearing a dirty cap. As always, when I came in contact with a black person on the white man's land, I braced myself, not certain who we would be with each other, whether there would be a nod of recognition and camaraderie. Sometimes there was coldness or contempt, or nothing at all, but there was always that instant of electric scanning to gauge who had the lighter skin and who the darker, who was more cultured and who more ignorant, a country "Tom."

I remembered the first day I'd encountered one of the mowers. His eyes clocked my green eyes and light skin before they flared and hardened, registering that I was black, as if to ask what I was doing on that side of the fence. The next week, when he chanced to bump into me, a furious curtain drew across his face.

Today I could risk no more bruising, so I turned back toward the dorm. But there in a flower bed was a little black dog from down the street. He had settled into the cool, moist dirt and was slowly and thoroughly licking his balls. I was surprised how much the sight shocked me, as if dwelling with the nuns had made my psyche permeable and I had absorbed their attitudes unawares. I looked around to see if anyone saw me and then called to him. He glanced up and wagged his tail so hard that his whole backside swung back and forth. He had happy brown eyes, and when he stood he was pigeon-toed. "Hello, sweetheart," I said, and he gamboled over.

~~~

I woke the next morning with the sense that I'd lain in the same rigid position all night, and a huge despair had paralyzed me. The iron bell began to toll morning prayers, but I didn't think I could stand to see a single one of the sisters, mumbling their canons about Jesus' love and God's mercy. I got out of bed and went into the bathroom. I ran a scalding tub of water, stripped, and climbed in. I eased down into the heat, watching the hair on my arms and legs raise.

After a time, I heard Sister Mary Richard's heels on the stairs, her swishing skirts, and finally the voice that had so recently given me such hope.

"Rita, are you in here?" she whispered, sounding alarmed.

"Yes, Sister," I replied.

"Aren't we a bit late bathing this morning?"

"I don't feel well this morning, Sister," I said. "In fact, I am going to go back to bed."

I could feel her sorting her thoughts. "What's wrong?" she asked at last.

"I don't know. Stomach, I guess, Sister," I said. It was the truth. Since we had spoken, my gut had been in a knot.

"Do you need to see a doctor?"

"I don't think so." I hung my toe in the faucet. I could feel her thinking, wondering what to do, but I knew she would never intrude on me in the bath. How peculiar it was to find safety in being stark naked.

"Well, you may go back to bed, but I am going to check on you at noon, and if you aren't better you are going to go straight to the doctor."

"Okay," I said. When she was gone, I contemplated the mountains of my knees rising above the sea of bathwater.

Five days passed, during which I barely ate or slept. I went to the library each day and dusted the shelves, but I was indifferent to the books that had been so beloved to me. I avoided even Sister Mary Cordelia. It appeared that no one was going to come and talk to me about any of what was causing me such pain. I would have to find out, in my own way, what being black might mean to me. What God might be. What a soul was. The dogma I had been handed was just not big enough.

On Saturday, after morning prayers and breakfast, I went back to chapel. I sat there in the prismed light, wondering about the man who owned the motel swimming pool. Was he from Alabama, with a fat gut and bad teeth, tobacco juice staining his beard, a rifle rack in his pickup? Or was he sleek, with too much aftershave? After a time, I lay down in the pew and dozed off. When I woke, it was nearly noon. But I was clear about what I had to do. I needed to go see that man and look him in the eye, find his truth.

I went to see Sister Mary Bernard and told her I had to go to town to buy some special shampoo for my hair. This happened to be true—it just wasn't the only truth.

She looked at me, then said, "Be back by vespers." I was shocked at how simple it was.

After lunch, with the sun so bright I could feel it cooking my scalp, I set out down the winding circular drive. When I glanced back, the school looked immaculate, manicured down to its purple pansies. It made my throat close.

About eight blocks down was the drugstore, where I got some bubble gum and a packet of bobby pins. I didn't bother with the shampoo because it would be burdensome to lug it all the way down the road. I asked the man behind the counter how to get to the motel. He looked at me over the rims of his glasses. "Down yonder. But it's a fur piece. You on foot?"

From the corner, I could see the bus station where Daisy and I had waited for the cab. Until the moment when I stepped off that corner and headed for the motel, I had done nothing wrong. Did I really have to look that man in the eye and find out for myself what the difference was between me and someone like Caitlin? The price could be steep. I could get kicked out of school. Daisy would never understand. I stepped off the curb.

The town looked bleached, as if the surrounding desert had leached all the color from it. Finally it just gave up, and nothing but scrub and the prison graced the landscape. I trudged along asphalt so hot it drained off the shoulder in gleaming black rivulets of tar. Cars raced by, kicking up dust. I thought about putting my thumb out for a ride, but now there were no more cars. In the distance, the bells of Holy Cross Abbey, where the Catholic boys went, tolled solemnly.

I sang horse songs to myself to drown out the voice of Daisy's alarm in my head.

> The Tennessee stud was long and lean
> The color of the sun and his eyes were green

The rhythm of the song gave me heart. It made me think about my daddy on his horse in Nebraska. He was a real Westerner, a real cowboy. And I was too, and I wasn't going to crawl.

I had almost talked myself into a state of courage when the long, low shape of what could only be a motel came into view. I could smell the dryer on the wind even though I could not hear it. Same scent as when Daisy took me with her to clean the Rabbit Ears Motel. I could almost picture the people who ran it. There was a constancy to the faces of people who worked in motels, worn out from being on call twenty-four hours a day. When it was busy they were tired, and when it wasn't busy they were scared about money. It just never let up.

An unlit neon sign depicting a lady in a swimming cap and suit had been nailed to a post high above the drive. The smell of chlorine grew stronger as I drew closer, as did my desire to turn back. I resolved at least to get a glass of water before I made any other decisions. I was chagrined to realize how grimy I was with dust and sweat. I was going to show up looking just like the very stinking, dirty, po'-ass Negro he didn't want. I could hear kids yelling and splashing in the pool behind a gate.

I pulled myself as tall as I could and strode up the driveway like I belonged there.

The lobby was painted in avocado, with a worn tangerine carpet and a honking air conditioner. The woman behind the counter had worked hard on her appearance. She had shellacked her black bubble do with hairspray I could smell all the way across the room, and tucked a couple of yellow velvet bows at the temple. Ultrafrosted lipstick made her lips look greasy, and clots of mascara weighed down her lashes. She was intent on painting her stubby chewed nails with pink polish.

She looked me up and down as she blew on her nails.

"Just so's you know, they just hired me, so I don't think they need no help."

"I need to speak to the owner," I said. I was shaking, and not just from the air conditioner.

"You're not looking for a job?"

I shook my head. She studied her thumb. "Okeydoke." She dialed the phone with a pen.

"Hello? Angelo in there?" I stared at the vending machines in the lobby, my mouth watering at the sight of so many peanuts, M&M's, jawbreakers, and gumballs. "There's a little kid out here wants to talk to him."

What did she mean, "little kid"? I was thirteen. Was this a racial thing, the way Uncle Billy was always talking about white people calling black men "boy"? Was there ever going to be a way for me to sort it out and tell when I was a human being with human beings and when I was the black problem? I was beginning to see how crazy it could make you never to be able to tell when they were toting a rope behind their backs. Was I an idiot to believe it would be enough just to ask?

The woman—she was a girl, really—looked up and, out of nowhere, she smiled, like the sun emerging from behind a cloud bank. This small kindness nearly broke me. "Angelo. He's down the hall in the bar."

I stood there on the carpet for a moment, listening to a fly bang up against the glass. It had never occurred to me the motel would have a bar. I couldn't go in there. If the nuns found out, I'd get kicked out of school for sure.

The girl, who had started on a second coat of polish, looked up at me once more with a questioning look. "The bar's down that way." She nodded with her head toward the swinging doors, and I turned and walked down the hall, toward the unmistakable stink of ferment and sour dishrags.

After the intense sunlight outside, I was blind in the darkness of the bar. The place was as cold as the locker in Steamboat where we stored the wild meat we couldn't fit in the freezer at home, but there was a slightly raunchy, intriguing quality in the air too. It reminded me of Mama and the burn of her whiskey-soaked cherry and the oily way the liquor coated the fancy glass. But if I could smell the cigarettes and beer this strongly, I had no doubt that Sister Mary Bernard would too, later on.

"He's a Rebel" was playing on the jukebox—"*He never ever does what he should.*"

"Whoooeee! Would you just take a look at that spring chicken," said a man I couldn't quite see. Gradually I made out red leatherette booths along the wall. On the other side of the bar four men were crouched on stools, studying me frankly over their beers. A couple of them wore crushed cowboy hats, one was bareheaded and bald, and a fourth was wearing a cap that said, PROTECTED BY SMITH & WESSON, just like Uncle Billy's. The fifth man was a businessman with his white shirtsleeves rolled up, his tie undone, drinking hard. He barely glanced in my direction.

A sixth man came down from the far end of the bar. He had allowed his shirt to hang open, revealing a gold chain sporting a scorpion that nestled in his black chest hair. He looked at me over his dark glasses. I wondered if he was wearing them to hide his red eyes, like my uncle Ernest did when he'd had too much Tokay. He stood on his toes to get a better look at me and then he whistled. "You too young to be working in here," he said, in a low whiskey voice that both alarmed and thrilled me.

All my horses scampered for the four corners of the room, and the lofty speeches I had rehearsed collided in a jumble. My fantasy of heroic confrontation paled under the glow of the Pabst Blue Ribbon sign. "Are you the proprietor?" I finally asked, staring at the floor. It needed vacuuming, I noticed.

He nodded. "You know you got to be twenty-one to drink in the state of Colorady, and that's doubly true right here in spittin distance of the pen."

"I don't want a drink," I said, and then I realized how parched I was. "I need to speak to the proprietor about the swimming pool."

"The swimming pool?" The men at the bar looked at one another. I wished I could talk to this guy alone.

"Your family staying here? I don't remember checking you all in."

"I'm from Saint Scholastica."

The jovial curiosity in the room downshifted as all six men swiveled on their stools to stare at me. "I want to talk to the man who told Sister Mary Richard I couldn't swim with the rest of the kids because . . ." I couldn't say the word. "Because I'm a Negro."

The man pulled down his glasses and snorted. "Naw. Get out. I mean . . . it don't show."

This completely threw me. Was it some trick—first humiliate me, make me less of a person because of the shade of my skin, then make a joke about it? No wonder Daisy was crazy. But I hadn't walked all this way and risked getting kicked out of school to let him weasel out.

"Well, I am black, and that certainly was the point when you told Sister Mary Richard you wouldn't let me swim here with the rest of the kids because of it."

He took a step back and cut his eyes down the bar at his patrons, who were leaning forward over their beer glasses, still as a herd of steers hearing a strange noise. Then he came out from behind the bar, wiping his hands on a rag.

"Jesus Christ, honey, you walked all the way here from the girls' school in this heat? That's nearly five miles."

He took off his sunglasses and looked at me with his hard eyes, as if he could strip me down to my white cotton panties. I felt woozy under his gaze. I hadn't seen a real man in months, and here I was in a room full of them. I had forgotten the hum of their engines, how they inhabited space, drew all the air and light into themselves so that only they existed. I was full of so many different emotions, they were about to explode from me in tears. I could feel them pushing against the back of my eyes. I knew I would never forgive myself if I broke down bawling like a little goat, so I willed myself to stare back, at the point between his eyes.

"Look, I just need to know if you said that." I couldn't say any more. Either he would talk or I'd just turn around and leave.

The jukebox was playing "Listen to the rhythm of the falling rain." But I willed myself not to be lulled by the melody. The trucks I could hear out on the highway, bringing peas and paper, were not running for me. Nothing was. My life was to be as arbitrary as Daisy's, at the mercy of other people's whims. My resolve in shreds, I began to sob.

"Oh, Jesus, kid, don't cry." He lowered his voice and hurried over between me and the other men so they couldn't see me or hear what he was saying. "Come on over here." He herded me over to the wall behind the cigarette machine. "Sweetheart, it weren't nothin personal."

I yelled at him then. "It is, too, personal! You personally said the person who is me isn't as good as any old stupid white kid."

He put his hand on the machine and leaned on it as if he needed strength, studying me with surprise. Finally he said, "It would be fine if it was just me. But most of my regulars in the summer come up from Texas and Oklahoma and I . . . I had to take out a second on this place and there's hardly no business in the winter, 'cept folks visitin at the prison."

Suddenly I was too tired to push this anymore. Nothing I could say was going to make much difference anyway. "Could you give me a glass of water and some of those cherries?"

"He absolutely can and will," said a bluesy voice behind me, riding in on a perfume barge. "Connie told me to get ovah heah right away." I turned to see who was speaking and encountered an eerie painted likeness of Elizabeth Taylor's face over frothy hot pink ruffles that struggled to contain a floating cleavage. "I'm Mrs. Cardalucci. Can I help you? My old man started a fine mess here, din't he?"

Angelo sprinted to the bar for water, looking grateful to have something to do.

"You poor baby. I told Angelo it was the wrong thing to do."

Something about her accent—not West, not South—undid me, and I started to cry outright. She put her arms around me and it was as if my entire body let go, turning itself inside out. "There, there, little lamb. Hurt your feelings din't it? I am so sorry. Angelo, let's get—what's your name?—let's get Rita a Coke," she called, steering me out of the bar.

I let her lead me, as if I was four. She had the big soft feel of a mother who understood. She guided me into a cheery dining room with turquoise vinyl booths and walls the shade of Aquamarine hand lotion. She picked up a napkin and began to mop the tears from my cheeks. "You came all the way down here from the school to talk to us? You must be starved." She called down the hall to the reception desk for sandwiches, then turned back to me. "Hope you like bologna," she said. I brightened. I hadn't had a bologna sandwich since Steamboat. I sank into the cool, smooth vinyl, realizing that it was the first time I had sat down in over three hours. A couple of blisters throbbed on my heels. The room soothed me, the light refracted through a glass-brick wall reminding me of the pool back home in Steamboat where I had learned to swim, my only question then whether I was going to drown.

Mrs. Cardalucci came back to the table, jingling. She was all ornament—hoops on her ears, gold charm bracelet, four chains struggling for purchase between her breasts, and smack in the middle of the crevice between them, a huge diamond cross that struck me as sacrilegious. I wondered where she got the nerve to go around like that. Daisy would have been scandalized.

But then, everything sexy carried more load if you were black. What logic did the race thing hinge on? Was it always random as a hawk diving down out of the sky? And so complicated? It was unnerving to have the enemy I had rehearsed confronting turn out to be this thick-bodied Italian and his fleshy, flashy wife.

"Hell*ooo*, anybody in there?" said Mrs. Cardalucci, snapping her fingers under my nose. I realized she had asked me how old I was.

"Thirteen years and one month."

"You got some guts."

The girl from reception brought the sandwiches. I gulped mine down while Mrs. Cardalucci nibbled at hers. Outside, trucks passed on the highway, rattling the windows.

"I'm not saying I know what it feels like to be Negro," she said, biting daintily into a cherry pepper. "Nobody can understand that, but you should have seen how they treated Italians in South Philly. We were in a neighborhood with Irish—everywhere you looked, a mick cop. And Angelo wanted to join the force. You think he could get a job?"

I hadn't seen anything but uniforms and habits for so long, I just let my eyes feast on her as she talked—the thick makeup, the lace at the cuff, the deep saturation of the pink dye of her blouse. I had never seen a more theatrical-looking woman except at Perry Mansfield, when I got the occasional glimpse of an actress putting on stage makeup. I could see that the aquamarine walls and the turquoise booths must have been her idea, as well as the hula-grass skirt for the hostess's little podium at the door. It struck me as strange that the sisters at Saint Scholastica had asked to use the swimming pool here.

"Mrs. Cardalucci," I blurted out, "are you guys Catholic?"

She looked at me, blue eyes fluttering in their mascara cage. "Of course," she said, as if it was obvious. "But that don't mean you gotta be a nun, you know." I must have looked stunned, because she said, "What, you think if you're not a sad sack you can't take Communion? Phooey on that, I say." How, I wondered, had she escaped the castration and branding that went along with being in the Roman Catholic herd?

"By the way, does Sister know you're here?"

I jumped up, icy fear seizing my stomach. "Oh, God. What time is it?" Through the glass bricks, the light was the color of mango. How could I not have noticed how late it was getting? There was no way I could get back in time for supper, let alone vespers.

"Don't tell me you just took off and walked over here without telling her where you were going."

"I said I was going to pick up some shampoo and that I'd be back in time for vespers."

Her mouth dropped open. Mine did too. They had probably called the pen by now.

"You could have been kidnapped by some prison runaway, right?" she said, giving voice to my fears. The word "prison" threw me into double panic.

"I'm gonna get Angelo. We got to get your fanny back right now." She started for the door, her spiky heels clattering as she ran.

It had been so nice to be away from the academy, not to have ice lining the walls of my stomach. Now it was back, so fiercely I could hardly move. By the time she returned, dragging Angelo, car keys in hand, I was bent over in terror.

"They're going to kill me," I said as we climbed into the front seat of their Cadillac, which smelled of cigar smoke. Angelo started the engine.

She tilted her head down at me. "Probably not."

I looked up at her.

"Honey, getting kicked out of Saint Scholastica is not going to kill you."

I was shaking. She put her arms around me again and hugged me hard, her gold chainery mashing my ear. "Shush, listen to me. What you did here is no small thing but that doesn't mean they are going to appreciate that."

"But what am I going to do?"

"Be yourself."

That didn't sound like a very good idea. I tried again. "But what am I going to do if they kick me out?"

"Nobody gets to know ahead of time."

And before I knew it, we were at the school, and all the lights were blazing.

8 ~ Bad Penny

My aunt Grace

~~~~ **The bus from Canon City** to Steamboat not only stopped in Denver but meandered through all the little towns along the way, giving me plenty of time to appreciate the expense of my victory at Mount Saint Scholastica. The two hundred miles back over the Continental Divide to the soaring western slope of the Gore Range would take all day.

At first only four passengers hopscotched along the cavernous expanse of the coach as it labored across the plains: a pregnant young mother and her gooey, drooling baby, a short Latin man in stiff new jeans rolled up at the cuff, a young soldier whose jarhead cut and stuck-out ears made him look like a kid playing army. When they scanned me, what did they see? An excruciatingly shy teenager with lopsided glasses, and a quartet of braids that exploded at the ends like stalks of broccoli.

I would have preferred to burrow in the back of the bus, but as luck would have it the driver was Danny, who stashed me in the seat directly behind him and kept up a steady patter about the towns we passed through. Florence, with its wide streets and towering grain elevator at the edge of the city limits. Pueblo, where they stuck the mentals. And finally, Denver, where we had a lunch break.

Inside the depot, migrant fruit pickers with their huge families, fragrant tortillas, and generous hearts mingled with college students in expensive hiking boots, soldiers and air force cadets, and earnest teams of Mormon boys in their bow ties, their eyes ablaze with mission behind their horn rims, while gaunt and wary cowboys surveyed the herd. There were black people too, adorned in teal and yellow, plum and scarlet—colors I hadn't seen anywhere in months. I thought about calling Uncle Billy and Aunt Helen, but devoted Catholics that they were, they wouldn't be pleased to hear I'd made a misstep with the nuns.

I longed to buy a *True Romance* or a *Police Gazette* to relieve the tedium of the rest of the trip, but Danny never let me out of his sight. He dragged me to the cafeteria and ordered me a burger and a glass of milk. "Not hungry? You will be. Drink the milk."

When the big rack of the Colorado Rockies finally came into view, I burst into tears. I was going home, even if Daisy didn't want me. Danny clicked on the microphone and resumed his running commentary of this stretch so familiar to me I could practically recite it. Golden, home of Coors Brewery. Central City, new opera house. The Continental Divide, average annual snowfall 315 inches. Fraser, coldest spot in the lower forty-eight. But even as Danny kept up the patter, I felt him keeping a wary eye trained on me in the rearview mirror as various cowboys and roughnecks sauntered aboard and appraised my breeding potential. Danny's uniform jacket stashed in the empty seat next to me made them hesitate before they moved on down the aisle.

In Kremmling, a showroom window full of flashy Buicks gave me a sudden pang. *Automatic automobiles,* I thought, picturing Sister Mary Richard at the blackboard during the last of our afternoon sessions. She had been emphatic that I learn both Latin and Greek. "Words come in families," she said. "If you know a word's root—its parent—you can infer meaning even when you don't know exactly what it's about." She wrote on the board A-U-T-O and underlined each letter. "*Auto* is the Greek word for 'directed from within,'" she'd said. Her veil swirled behind her as she turned back to me. "*Author, authority, authenticity, authoritarian.*" And when she began to explain the words, I then saw what she meant. *Authentikos* was the Greek. *Authenticus* the Latin. But it was the next word she wrote down that pierced me.

Autonomy, governed from within. It was at the core of everything Daisy was driving me to do on her best days, and on her worst, the very last thing she'd allow. Why would God give me a mind if it wasn't to be used? Wasn't the very kernel of slavery lack of autonomy? Why would Daisy, who bore the wounds of slavery, slavishly embrace the most enslaving ideas of the Church?

When I was about nine, the first pair of visiting nuns had

come to Steamboat in the summer to teach the local Catholic children catechism. Their black habits shocked me. The nuns who had kept me when Mama first died wore white. I knew that black was the color of evil. In the movies, the bad cowboy wore the black hat and rode the black horse. Why would the priests and sisters choose such a color?

What made these women holy, I wondered? Having cowboy boots didn't make you a cowboy. I had cowboy boots. And Slim McCormack, who was a real cowboy, with his bowed legs and sun-squinted, gentle blue eyes, sometimes wore work boots. No matter what he wore, he could never *not* be a cowboy.

But what scared me even more was how cringingly obsequious Daisy was to them. Before they went back to their motherhouse that first summer, she invited the sisters and the priest to our place for a supper of snowy biscuits, shortcake, freshly butchered pullets, and the last of the chokecherry wine. "Now you stay in the back until I call for you," she told me. I wondered why I couldn't eat with them. But then I saw she wouldn't sit down with them either, standing and serving them the whole time. And even though they protested, she remained planted behind the counter near the stove the whole time, wearing that red bandanna.

I watched them from the back of the house. They buttered their biscuits just like ordinary folks. Yet who they were didn't matter any more than who I was mattered. This costume was all they needed.

The following summer, in a fit of devotion, Daisy promised Father Funk fresh berries all summer, guaranteeing our poverty. She sent me scrambling up Bear Creek to gather thimbleberries and sarvis berries between the periods when our bushes weren't setting fruit. I didn't mind the chore. I liked hurrying up the hill just as the night shift—skunks, raccoons, mice, and owls—

were returning to their burrows, while the day shift—magpies, horses, cattle, and humans—awoke. The early morning sun revealed soft hollows near the willows where deer had rested in the night.

But when we brought the berries to the church, I wished Daisy would just drop them off at the rectory rather than following me down the basement into the classroom where I knew my friends from school would already be at their little desks. However, she'd clopped down the stairs with the flat of fruit, pushing me ahead of her.

The nun had drawn three milk bottles on the board. The black one was labeled "mortal," the gray one "venial," and the white one "pure." Something about this bothered me, but I was all wrapped up in trying to decide whether to join the other children or to hide outside the room until Daisy finished. "The state of our soul when we are free of sin is white," the sister was saying.

But Daisy's zeal to complete her mission carried her right into the classroom. "Excuse me, Sister," she said. The nun turned around, startled.

"Sister, my name's Daisy Leonard, and I'm a convert. We picked these fresh for you this morning." Daisy looked around and noticed I was still in the hallway. "Rita Ann, come here." Reluctantly, I stepped forward. "This here's my niece I'm raising because her mother's dead." I could feel ten little sets of eyes behind me, taking it all in. Just because they saw me at school didn't mean they knew Daisy.

The nun accepted the berries and smiled politely, looking around for a place to put them. "Don't these look delicious," she said. "Thank you—"

Daisy cut her off. "She ain't had no kin since she was four and a half. Been raisin her to be a good clean Catholic."

I studied the floor furiously, my eyes riveted to the glossy painted concrete.

"Did you grow these yourself, Mrs. Leonard?" The nun struggled to make a place for the berries among the papers on her desk, but Daisy didn't seem to notice.

"I converted my whole family. You can believe that too," she said, warming to her story. "Every Saturday night, we sit in the dark and say the rosary. And then after that we have a wash-tub full of popcorn."

I heard a snicker. If God was so merciful, why was this happening?

The sister turned to me, still clutching the flat of berries. "And what is your name, young lady?"

Like a calf being presented at the county fair, I had not ex-pected to be addressed. "Rita," I whispered.

Daisy thumped me on the head, breaking my reverie. "Quit your damn mumbling and tell Sister your name is Rita Ann." She laughed self-consciously. "Now look, you made me cuss." Behind me, the rustling and tittering were growing.

Daisy tilted her head and her face contorted for a moment as if she was about to cry, giving me a moment of panic. "I been through nothin but tragedies. We lost her mother, my sister. And my mama and papa and I raised her sister, and now I got her. If it's not one thing, it's another. Now she's got to have glasses."

I could feel my ears burning deep red.

"Well, Mrs. Leonard, thank you for the fruit," said the sis-ter, setting the berries on top of a steel radiator by the window.

I knew Daisy wasn't going to let this city slicker ruin them. "You need to keep that fruit cold if you want a good shortcake tonight."

"Of course," said the nun. "Thanks again."

Daisy hesitated before she headed for the stairs, torn between fetching the berries to a fridge and acknowledging the pain of her dismissal, which showed so plainly on her face. When I finally took my place among the other children, I was far beyond mortal and venial. Try as I might, I could not imagine how to pour every thought and deed into three milk bottles—black, white, and half white. Or was it half black?

~~~

The sight of the goofy Rabbit Ears Motel sign broke through my fretful memories. I didn't see it quite the same way after my adventure in Canon City. Even though it was only four in the afternoon, the NO VACANCY sign was turned on. Full already. That was a good sign. I couldn't help being thrilled by the sight of my bustling hometown. Steamboat really was a prize in the summer, when the surrounding mountains peered down at the town like it was a pretty newborn in a green bassinet.

Daisy was waiting under the eaves of the Harbor Hotel, looking, as always, like a refugee from another century. She was wearing her Conestoga bonnet, jeans, and the rubber overshoes with big black buckles that meant she'd just come from working in the yard. But thrown together as she looked, there was nothing haphazard in her stance, arms crossed, eyes squinting against the slanting rays of the sun just before it tucked itself behind Storm Mountain. She scanned the bus windows like a bear who had caught a worrisome scent.

When the bus door opened, the air bit surprisingly cold. I had forgotten how much cooler summer was in the mountains. Danny positioned himself by the door and smiled broadly when he saw Daisy. "Brought your young lady home," he said, tipping his cap to her and offering his hand to me. When our eyes met, Daisy sniffed at me, but she didn't say anything.

"How's Mary getting along?" Danny asked, still holding on to me. Why did he have to bring up the vanished favorite just when this bad penny had made her way back?

"Not a word. Any snow left on the pass?" Daisy said, heading him off.

Danny registered the rebuff and abruptly strode toward the luggage compartment. I retrieved my twine-wrapped box and trod off after Daisy, who still hadn't spoken to me.

Tourist season was in high gear, and the little Jeep, with its homemade wooden cab and fading army-green paint, looked especially out of place flanked by big new cars with out-of-town plates and plenty of shiny chrome. I put the box in the back and took one last breath before I got in.

Daisy didn't start the engine. She just looked at me, not breathing, until I felt the skin on my forehead begin to prickle. I studied a spot where the floorboards had rusted through, letting in a glimpse of the pavement below. I wondered what kind of punishment she had planned for me this time. Would she take me straight to the sheriff and have him drop-kick me to the reformatory, or did the separation of church and state prohibit that?

"Your head looks like a nickel's worth of bat shit," she said blandly, flexing her fingers on the sun-baked steering wheel.

"At least let me tell you what happened," I began.

"I don't give a goose's hind end for what you say happened. I *know* what happened. What happened was you done runned off and bitched up a chance at a good education and I still got to get down on my knees and scrub toilets to pay them sisters for your tuition, room, and board. *That's* what happened." She started the engine and shifted into first, popping the clutch so hard that the Jeep jerked out of the parking space.

I was just wondering whether to say I'd work all summer to pay back some of the debt when she plowed on. "Well, you can just plan on cleaning houses all summer. Maybe that will make you think the next time—if there *is* a next time."

Cars sped past us on Lincoln, the main street, as Daisy inched along. Sunburned tourists peered blandly into shops selling turquoise trinkets and horseshoe-shaped souvenirs. It was as if their tribe had been allotted some peculiar freedom that eluded us. *Autonomy.* My getting thrown out of Mount Saint Scholastica represented not just my personal failings, but yet another failure to fulfill the unspoken mission that held us both hostage: the redemption of all her losses.

We passed the red brick high school with the big blue spruce that had seen Rose Marie and Mary on their way to classes, band practice, and PE. I knew it would not see me through to the end.

"Daisy, I took a stand on something important," I began again, but I had not so much as gotten the words out when she smacked the steering wheel with the back of her hand. "What in Jesus' name could be more important than you gettin a good education?"

"But they were wrong—" I tried.

"Who cares if you righteous if you spend your days at the ass-end of a mop? Don't nobody care what an ignorant Negro thinks."

We had crested the top of a hill, and Daisy downshifted abruptly as three riders came into view, astride huge bays fitted out with expensive tack. They sat their mounts with the easy confidence that bespoke layers of training, experience, connections, resources. Daisy smiled widely and waved at them, giving them careful berth. Then she continued. "I'm done raising other folks' children. You turn eighteen, you're gone." I turned and caught the sudden look of fury on her face. "And don't think you got some place to ease off to if you come up big."

She didn't speak again until we turned onto Bear Creek Road. "Your Aunt Gracie's here from down South. Got here a couple days ago. William and Ernest come too, so you gon' have to bunk in the basement."

The only completely predictable thing about my clan was that they never followed any predictable pattern. I perked up. With company present, my punishment wasn't likely to include Daisy whacking me with the berry pail belt. I was relieved, for her sake as well as my own—I was no longer certain what I would do if she hit me again. And I was intrigued. The uncles were no particular surprise, but I had never met Grace, and never expected to.

The foliage around the house was at its most resplendent, as if it was completely aware of the adage, "Make hay while the sun shines." The blue spruce that towered over the house had borne pale clumps of new needles, and the creek bank in front of the well house was adorned with fist-sized, ballet-pink peonies, cobalt irises, rosy sweet william, tiger lilies, poppies, and night-blooming jasmine. In the yard, the raspberry canes had nearly doubled their growth. And over by the berry shed lounged three black people, all of them so willowy long that their legs had to be stashed sideways under the table.

Daisy frowned as she drove the Jeep across the bridge and into the driveway. "You'd think with three grown adults sittin up in that yard, somebody would think to lock the gate."

"Well maybe they think with three adults in the yard, the gate doesn't need to need to be locked." It popped out before I could yank it back.

Daisy blinked twice, as if her eyes stung, then her voice went very soft. "Rita Ann Williams, I am just two shakes of a lamb's tail shy of turning right back around and handing your sorry ass over to the sheriff. Now straighten up." She jerked the key out of the ignition, opened the door, and was out of the Jeep in one fluid movement.

"Gracie," she called. "I brought Mae's child."

I sat forward, to get a good look at this sister Daisy scorned for marrying and settling down in the backward South. She suspected Grace of being a follower of the snake Christians, and a bit simple as well. Still, she always said that Gracie had been the family favorite. "She was the only one of us could get Papa to laugh. She'd act crazy and Papa would just drop the belt." Watching her now, I could see why. Lanky and fluid, Grace was aptly named. Just now she had given herself over to some joke between the three, flinging her long arms in the air and laughing as if to throw off the thing that possessed her.

When I got out of the Jeep, Jiggs began to howl like a wolf. The sight of him clutched my heart. He looked like a hobo, his winter coat hanging in matted hanks, the heavy chain around his neck wound in fist-sized clumps that had shortened it so that he could reach only a few feet past the porch. Everywhere else in this country, dogs ran free on their owners' property. I stuffed down my anger, though, knowing full well that bringing up his condition would only guarantee him a heavier chain. Instead I ran and hugged him so hard he yelped, and then I began to twirl him around, trying to unclump the steel links.

"Ain't that just like this little wild Indian to run to the dog and not give the adults their propers," said Daisy.

I rose and went to greet them. Gracie stood to meet me, bending down to kiss my forehead like a giraffe. Then she held me by the shoulders at arm's length. "I do declare, Rita Ann sure is a little pretty thing. With all that good hair." She tugged at one of my braids. I knew what she meant by "good hair," that it could be made to look more like white hair. Her own was short, pressed, and gray at the temples.

"Something about her brings Chesley to mind." She had the voice of a Southern belle, to go with her mauve flowered dress.

I had once overheard a veiled exchange between Ernest and Daisy about Chesley stealing some peanuts from a white man's shop, but like their phone conversations, it seemed to be spoken in code. I knew he was one of the brothers who had not survived. Had he been killed? Starved? But before I could say anything else, Uncle Billy pushed himself up from his chair to give me an abrupt sideways hug.

"Hello, hello, hello." Uncle Billy always repeated things for emphasis. Of the four, he was the shortest, with skin the color of strong tea and muscular arms chiseled from lifting heavy pots on the Union Pacific railcars where he worked as a cook. He was also the most Indian-looking of the four, with a soaring beak of a nose that flattened out at the bottom as if it had remembered Africa at the last minute. Like my father, he had never gotten past the third grade, but he owned a tidy brick home in Denver, full of shiny appliances and the French pink furniture his Louisiana wife favored. Uncle Billy did everything fast, like he had a train to catch. Now he rattled off one question after another, ending on, "Heard Daisy sent you down to Canon City with the sisters? Did they learn you to cipher good? Let me hear you count." I didn't want to say that, working my hardest, I had only managed Cs, but I couldn't think what else to say. "What's the matter with her, standin there lookin?" Billy demanded, and as usual, I went blank and lapsed into defeated silence.

"She can count whatever she has a mind to," Ernest said. "You dig?" He scraped a kitchen match across the underside of the table and lit a Pall Mall. Then he choked on the smoke and settled into the wet wracking cough that was the sound I'd always thought of when Daisy mentioned his name. When it passed, he took another drag, and this time he managed to blow out a blue-gray cone of smoke.

Ernest was a peculiar combination of obscure and flashy. He preferred to peer out at the world through the double protection of dark glasses and a hat brim. Today it was a fawn bowler. Beyond the wine that was always on his breath, Ernest had no scent. Quite literally, he was cool. Rumors attached themselves to his tall obscurity like burrs. That he had owned several properties in Steamboat but lost them all, one when some town roughs burned him out, another the casualty of a card game. The wildest tale concerned him managing a turkey farm when a tractor-trailer arrived to load up a thousand birds for the Denver stock show. Supposedly Ernest claimed they were his, and pocketed the money.

Grace cupped my chin in her palm and studied me. "She got her daddy's eyes."

"Let's hope to hell she don't have his sight," said Daisy. Constitutionally incapable of standing still, she had gathered a handful of raspberries and was now absently plucking the deformed ones aside.

"Oh, Daisy, now you go on," Grace said, "Lee wan't but so bad. Mr. Anderson set a big store by him." I wondered how much of the goods she had on the whole clan. "How old is she now, goin' on fourteen, you say?"

"Goin' on twenty-one," said Daisy.

"She get enough to eat? She looks puny," said Grace.

I began to squirm. Had the slaves in the old days been inspected and discussed like this? Besides, what I ate was always a sore spot. Why did she have to bring that up? But to my intense surprise, Daisy was looking at me with a glimmer of pride in her eyes.

"She may be a vinegary little pisser, but she a Graham, all

right." Abruptly she turned back to her siblings. "So what all was making you three laugh so hard when we drove up?"

At this, they all broke up again. "Gracie'll spill the beans," Daisy said, ramming her hands down on her hips like she was about to spank somebody. Yet underneath her severity there was a twinkle.

Grace sneaked a look at Uncle Billy. "We was thinking 'bout that day when that preacher come and while his eyes was closed askin the blessing, you stole that chicken wing off the plate right under his nose, and the whole rest of the meal you could see him lookin round at everybody's plate trying to locate it."

Daisy's eyes narrowed and she crossed her arms. "Y'all ain't nothin. I'm fixin to go rustle up some Chinese checkers so you children can have something to occupy your simple minds." They all burst out laughing again as she stomped off, swinging her arms in exaggerated indignation.

"He was the only fat Negro I ever saw down South," said Uncle Billy.

Ernest stretched out long. "Always looked constipated, if you ask me." Billy and Grace snorted with laughter. "Had a face like a hog been drug through a knothole backwards." I had never seen Ernest this funny.

Uncle Billy turned to me. "We hardly ever got meat. But every time one of the old hens quit laying and Mama'd get set to fry her up, damn if he didn't come draggin his sorry ass down the road."

It was on the tip of my tongue to ask, "What was my mama like?" when Uncle Billy stood up abruptly and brushed off his pants. "Let's go help Daisy get supper on."

At the bottom of the stairs, Grace put her arm around my shoulder. "You know, your mama was the baby, just like you." Her long, tapered fingers reminded me of Mary's.

"Rita Ann, run pull up some lettuce leaves and garlic," Daisy called from inside the house. I was relieved to have a chance to catch my breath. Down by the goose pen, the lettuce needed thinning, and I tasted a leaf. It was ruffled like green tulle, so much thinner than the iceberg served at the academy but twice as flavorful. The garlic, a new strain Ernest had introduced to us, was so pungent I could smell it even in the ground. An owl in the distance hooted at the approach of evening. I remembered how suddenly night fell at the academy. Now it would be time for Sister Mary Cordelia's lonely tea.

When I came back inside, Billy was helping Daisy with dinner. "Hand me them garlic tops," he said when I laid the creek-washed greens on the chopping board. Daisy plucked pieces of raw chicken out of their buttermilk bath to dust in a well of flour on the counter. Behind her, several pieces were already bubbling along nicely in the Dutch oven. Warming at the edge of the stove was cornbread and an enameled pot of cabbage with smoky ham. Even Ernest got involved, prying open a jar of watermelon pickle we'd put up last summer and whipping cream for the raspberry-rhubarb cobbler Daisy had made.

"Somebody bring me a vase," called Grace, coming up the steps. I could hardly see her behind a pitcher full of frothy peonies and salmon poppies.

"If you want 'em to last, light the ends a-fire," said Ernest, backing up against the freezer as Billy reached past him for a jot of vinegar for the salad.

I got the leaf from behind the piano to widen the oak table, laid a tablecloth down, and placed the flowers and Mama's cactus-shaped salt and pepper shakers over a hole just off the center. I tried to imagine how she would have fit on this team of bustling brothers and sisters. How different they were together! Even Ernest's face had grown handsome as his big hands measured out a quarter teaspoon of vanilla for the whipped cream. This is what communion is, I thought to myself. Community. There it was again—those word families.

Judging by dinner that night, you'd never know Daisy had ever resorted to punishing deer meat and potatoes into boiled mush. She did it right for her uneasy family. The salty crunch of the fried chicken gave way to moist juiciness inside, the cabbage was both sweet and spicy, the grainy cornbread had been slathered with maple butter, the cobbler just tart enough to welcome the cream. And all of it crunchier, juicier, spicier, sweeter, because this was the kind of feast they dreamed of when they were starving. "Even Ernest ate a full plate," laughed Daisy afterward, "and he don't believe in food." Ernest reached into his shirt pocket and pulled out a Pall Mall, but the flicker of a smile appeared at the edge of his mouth.

"All right, kid, go strut your stuff on that piano," said Daisy, as she stacked the last of the dishes. My mouth went dry. It seemed impossible that my music had made the journey home with me. My aunt and uncles looked at me expectantly, but I could not imagine playing for them, certainly not the old stuff Daisy loved. I wasn't even sure I could remember how to play "Chapel Chimes." But I didn't exactly want to share Bach with Daisy. He was still barely my own.

And then Grace said, "Oh, honey, I bet you musical like Lee was. Did you know he built his own guitar and taught hisself to play?" It was enough to persuade me to try a Bach prelude on her. I waded in softly at first, aware that the piano lacked the rich timbre of the instruments at the academy. I had practiced the piece enough, though, that mechanical habit took over. The music gathered momentum until it felt like the creek at its highest spring flush. A surprised hush fell on my uncles and aunts, and I could feel that the music had gathered them up and brought them along with me. When I struck the final note, nobody said anything. Then I dared a peek at Daisy. She was smiling, and Billy and Gracie were openmouthed. Even Ernest was looking directly at me. "If only Mae could have heard that," Grace said, stroking the tablecloth.

~ ~

I woke the next morning to the rough basement walls, the smell of rich coffee, and a confusing sense of well-being. I had slept until nearly eight. I came upstairs to a homey clean kitchen. It was so nice that someone other than me had finished up the dishes. Overnight, the peonies had blown wide, and their perfume lay lightly in the bright room. Jiggs heard me moving and called to me in his familiar half growl, half yawn. I smeared a slab of leftover cornbread with maple butter and piled scraps into a pail for his breakfast.

With a mouth full of sweet crunch, I went outside. The fat clouds brooding over Swigert's meadow had kept the sun from drying the cool dew on the garden leaves. The deer- and horseflies that usually tortured us by midday hadn't come out yet. Jiggs came to me, head low, dragging his knotted chain, bluff growling, wagging his tail so hard it nearly knocked him over. "Where's your brush, Dogman?" I asked roughing his muzzle. He licked my face and shook his neck. Would I unlock the chain and let him off today? he asked silently. I kissed him between the eyes and shook my head. Not with the relatives around.

Along the creek, wearing identical straw hats and holding fishing poles, Billy and Ernest stood as motionless as if they had been captured in a landscape painting. Grace, who had been bent over in the pea patch, unfolded herself, pressed her hands to the small of her back, and stretched. "Morning, Miss Music," she said.

"Howdy, Aint Gracie," I replied, doing my best to speak Southern to her and making her laugh outright. She had on my yellow flowered sunbonnet, made in the style of what settler women wore driving the Conestoga wagons west. I would have preferred a Stetson, but Daisy made the bonnets herself, folding the cloth and stitching it into sections that she stuffed with little cardboard slats to keep the bonnet stiff. She was always after me to cover up, lest I end up "black as a damn crow."

"Don't you fish?" I asked Grace.

"Not if I can he'p it," she replied. "I can't stand to see them poor little ol' eyes starin up, askin me why I done them that-a-way."

She told me Daisy had left for Perry Mansfield already. Apparently she'd decided to give me a day of grace before I got down to the cleaning business with her for the rest of the summer. Grace brought the basket of peas up onto the porch, breathless from the altitude. Even visitors from mile-high Denver felt it. "You reckon you can bring some chairs out here and help me shell these?" she asked.

I did as she asked, and before long we'd settled into a rhythm together. Grace worked quickly, deftly stripping the zipperlike strings from the pods, popping open the packets with her long fingers, and scraping out the dainty peas with her thumbnail. I knew that Daisy would hate the sight of us. Despite her bandanna and faded jeans, she had a special disdain for blacks down South, "settin up starin down at the road, waitin for somethin to happen."

"Aunt Gracie, what was Mama like?"

Grace studied me for a minute. "Well, she was quiet and sweet, like our grandmother—your great-grandma," she said, clarifying. "When the Klan boys got to stirring up a pot of mess, Daisy used to hide her with our grandparents. You see, they was half Cherokee and knew how to do."

The only Indians I knew about lived in the West in tepees, and Robert Ball Anderson had hunted them. Which prompted another question. "How could Daisy have married an Indian hunter?"

The resentment in my voice snapped Grace back from her reverie, and for a minute I thought she was going to get mad at me. But she said, "Sweetheart, do you know what it means to be a slave?"

I decided to answer a question with a question. "Aunt Grace, how come you never left the South? Daisy said it was a terrible place. Koo Klukkers and crazy snake religions and people taking cow piss for medicine and voodoo. How can you stand it?" I thought of the South as a place where blacks and whites were locked together in hatred, like mating dogs.

She said evenly, apparently taking no offense, "Well, baby, ain't no snakes never done me no harm. And after I married Mr. Carpenter, well, I wasn't gon' take and move up here jus' cuz Daisy said so. Besides, Aunt Gracie's bones don't take to no cold weather."

In front of us, the clouds were piling up beyond the immaculate, weed-free rows of vegetables. "How come Daisy gets so scared by a little bit of rain?" I asked.

Again, my tone carried more of an edge than I'd intended, and Grace said a little sharply, "You best mind your uppity ways, cuz you don't know what all Daisy been th'ough." She looked down at the dishpan, sifting handfuls of shelled peas through her fingers like little green pearls. Then she looked out across the pasture as if seeing something from a long way off. "You know she got struck by lightning, don't you? Twicet."

"I had no idea she'd been struck once," I replied.

"Well, folks don't like to dwell on all that's past," Grace said, "but I'll tell you about the first time." She popped a handful of peas in her mouth.

"Precious blood of the Lamb, it was jus' the evilest day. Teacher didn't show up to that one-room schoolhouse but four months out of the year, and school was set to start that very day. Durn if Pappy didn't tell us first thing that morning, 'Man needs his cotton picked, so y'all can't go to school today.'

"So we got out in the field, draggin them heavy cotton sacks, weevils and chiggers all creepin around. The air turned gray as a slug, closed up thick and sweaty." Grace shivered, remembering. "A person couldn't hardly catch they breath and there was no place to go. The land was flat as a plank and heah that ol' funnel come snakin down out of the sky like a cottonmouth." She looked at me and shook her head. "But Pappy took our hands and told us to hold on to each other and we runned and jumped down in a ditch and covered our heads. The tornado lit out in the other direction and tore that white man's cotton patch to pieces. And when it had a-plenty, it climbed back up in the belly of that black cloud and went on 'bout its business.

"Well, we reckoned it was all done with, so we stood up to brush the dirt and splinters out of our hair when this great big spike of lightnin reached down and struck Daisy so hard it must have knocked her back twenty feet. Like to killed her. We all runned up on her and so help me Jesus she was smokin. You know lightning likes to run through you and down in the dirt? Burnt a hole in her hands on the way out."

"Like stigmata," I said. Grace stared at me blankly. "You know, like when Jesus was nailed on the cross?"

"Jesus wa'n't nowhere round that day," Grace said. "Daisy was laid up a good long while. Loosened up her teeth, and her hair didn't grow good after that. But damn if she didn't raise up one night, light the kerosene lantern, and go on back to readin that book she had. Said she was still gon to be a schoolteacher."

"That was how she met Mr. Anderson," I said. I had read that far in his story.

"Baby, that's exactly right," said Grace. "Mr. Anderson had always wanted to learn to read and write, and that was the thing about her that he was most partial to."

"What about the second time?" I asked. I'd fallen into a rhythm now, shelling peas to the accompaniment of her voice.

Gracie looked out toward the creek as if to check that Ernest and Billy were still fishing, then dropped her voice as if it might carry all the way to them. "They don't like to talk none about this, but you may as well know. All three of them was in eastern Nebraska with Mr. Anderson when the same kind of bad weather come up. They said Ernest was driving, and you know how he likes to go fast. Daisy seen a funnel cloud and went flat crazy, and Ernest runned the car in a ditch. Well, Mr. Anderson was thrown out and they didn't have ambulances running to help black folks back in the day. So they picked him up and tried to carry him, and must have broke his neck.

"Well, you know how white folks love to talk bad about black folks . . . especially when Daisy came into that big ranch and all that money." She paused as a crack of thunder like a gunshot rippled across the valley.

"What about the lightning?" I asked, but Grace's voice had gone silent. I looked up to see Ernest and Billy running toward us under the gathering storm.

At the sight of her brothers, Grace zipped up like one of her pea pods. Although she continued to treat me in her gentle way that was as much as I got out of Grace for the rest of the visit. Perhaps Ernest and Billy said something, or maybe Grace simply felt she'd said too much. That night at supper, I learned that the three of them would be leaving in the morning.

When Grace saw the stricken the look in my eyes, she made a joke. "Well, sweetheart, you know how it is. After three days, fish and relatives start to stink."

~~~

The same nightmare that had plagued me after Mama died came back that night. A giant tarantula loomed over the rim of the mountain and panicked the trapped horses milling in the corral below. I struggled up yelling out of the dream to find Daisy standing over me. "Rita! Damn it. Wake the hell up." I swam up to consciousness, sluggish, but with a gut full of ice. She peered down at me over the rim of the flashlight. "Time to go." It was still dark when all five sets of knees, plus baggage, were finally sardined into the tiny Jeep.

At the sight of the Trailways, the brothers and sisters busied themselves with the mechanics of disentanglement. Uncle Billy ran everywhere, unloading suitcases, chomping down one last time on the fat cigar Helen would never let him smoke in the house. He grabbed my shoulders like I was the last thing to set in order and gave them a rough shake. "You get something in that hard head now, hear? You get that job out to camp and keep it." I nodded my head up and down, like a counting horse.

The darkness of Ernest's complexion made his mother-of-pearl sunglasses shimmer in the early dawn—big square cubes designed to mask his bloodshot eyes. I knew he was far too cool a cat to tolerate a hug from me. He was also fairly well-lit already, judging from the fumes I could smell even at my careful distance. Later, I knew, I would find a bushel basket of empty bottles out behind the shed.

High-heeled Grace towered above them all, a little loony in her tilted maroon hat and fox stole draped around her shoulders so that its little mashed face looked you straight in the eye. She grasped my chin in her hands and kissed me on the forehead, smelling of lipstick and vanilla. "Baby, soon as I get back home, I am going to send you a proper picture of Mae Ella." Then she stepped up to Daisy and gave her a hug. Daisy submitted, with her hands hung at her sides as if she didn't know what to do with them. Then Billy and Ernest picked up their "grips" and lined up to get their tickets stamped. I hated the hot tears that rose to my eyes as Grace's slender ankle disappeared into the darkness of the bus, my uncles right behind her.

As soon as the bus doors suctioned shut, Daisy muttered, "Simple as a Christmas goose. All these years gone by and I still got to send her a bus ticket."

I would miss them all, though I doubted I would see Grace again. Tender as I felt toward her, I knew I would never visit the land where white mistresses had once left their chamber pots in a little cabinet set into the wall, so the Negro servant could whisk it out of sight on the other. One thing I knew for certain: People never forgave you for seeing their shit.

9 ~ If the Creek Don't Rise

*Portia Mansfield dancing in the
early days of Perry Mansfield*

~~~~ **The entrance to** Perry Mansfield didn't look
like much. But for the plain board sign atop a lodge-pole pine,
it could have been the dirt driveway to any modest ranch in the
Yampa Valley, with the corral on the left and cairns of manure
here and there. Yet every time I passed through the front gate,
the same sense of yearning always overtook me. For what, exactly, I didn't know—money, a state of mind, a chance to mingle
with people of talent, grace, and deeply layered education,

people who dwelt upon the way light fell on the arch of a foot, or the density of shadow under a pointed toe. The place provoked some incoherent hope and hunger in me so profound that it made my soul ache, and I wanted to jump out of the Jeep and run in there and never come back.

Daisy leaned forward over the steering wheel. "Kingo called to say the folks they was expecting to help today had a wreck, so I got to go down to Cabeen. I'm going to put you out at the Julie and you'll have to give it a hit and a lick yourself. Sweep and mop the foyer first because Ingrid's got class at ten. Then do the ashes in the fireplace before you dust, else you'll just have to do it all over again. And don't forget to rub the windowsills with lemon oil."

The Julie Harris Theater was like a temple to me, and normally I would have treasured the chance to inhabit it by myself. But having to clean it was like having to dust the Stations of the Cross; it had about it the same air of sacrilege. But more than that, servants were never part of the show—their task was to support performers. Daisy was well aware that I was less than delighted to be cleaning alongside her this summer. She thought I had grown "uppity," believing I was too good for work. But I had seen how easy it was for a person to get defined by a simple thing. When one of the McMahon girls got her ears pierced, she was "slutty," and nothing she did after that could shake the label. Once I was seen as a cleaning woman, would I ever be thought of as anything else?

Daisy pulled up in front of the handsome low rose-tone building, tucked away under its conical steel roof. "All right, get your hind end moving," she prodded. I got out slowly. "And for sweet Jesus' sake, don't meddle with nothing having to do with they damn plays," she called after me through the window before she drove off. "I'll be back in an hour and then we'll both have to tackle Miss Marjorie's."

I was glad the foyer was empty at this hour. I wasn't ready to confront the usual gaggle of dancers in leotards, tight chignons gleaming, flexing and pointing their toes, working on their turnouts. The darkened theater smelled of dust and the water-based paint they used on the flats. I flicked on the overheads. It was odd to see the stage lit for work, the gelled lights dark. The pulleys and steel cables and weights to the fly space, the architecture of the set contrasted harshly with the scene you would see from the audience. I'd heard that they were doing scenes from *Tartuffe* this season, but in the dull morning light, it seemed especially hard to believe anyone would buy the illusion that this was a French villa.

I made my way past the set to the women's dressing room. In my haste, I stubbed my toe on a sandbag and nearly went flying into the property table. A garish papier-mâché parrot in a cage seemed to reproach me. And then I opened the door. Along one wall, costumes glittered. Heavy brocades and rich silks in wine and forest green, black dense velvets with golden cord. I reached out to touch a tunic and was surprised to feel how hard the stays were underneath the plush nap of the bodice. It had a makeup stain at the throat and carried a pungent smell of sweat. I slipped it on. Through a half-shuttered window, light flowed in gently as if loath to wake someone sleeping. I looked up and was surprised to see my own face in a mirror that spanned nearly fifteen feet. I moved closer and settled down on the bench in front of it.

Pencils and tubes and sticks and pots of makeup were arrayed in all directions. I picked up a stubby brush and a stick of bright blue greasepaint and outlined my lips, then filled them in.

Then I picked up a tube of flesh-colored paint and dabbled some above my right eye, some above my left. I hunched my shoulders into a Caliban hump.

> *You taught me language; and my profit on't*
> *Is, I know how to curse. The red plague rid you*
> *For learning me your language!*

My voice sounded squeezed in the little dressing room, so I returned to the stage. I looked out into the cave of the theater. It seemed to be holding its breath.

> *When thou camest first*
> *Thou strokedst me and madest much of me, wouldst give me*
> *Water with berries in't*

I was just beginning to warm to the sound of my voice in that vast empty space when something creaked out there in the gloom. I froze. I scurried back to the dressing room and I put the makeup tubes precisely back where I'd found them, grabbed a gob of cold cream and a tissue, and wiped my face clean, then took off the tunic, hung it up, and bolted from the dressing room. There was no doubt about it—when it worked, theater appeared to be magic. But it wasn't. It was makeup and lines, costume and sets, crafts that could be learned. If only I could learn them.

I set to cleaning with a will. I climbed in under the cone-shaped chimney to collect the ashes. Then I took a broom to the V-shaped foyer and dusted the benches that lined the walls. I did the bathrooms last, trying to bury myself in the scrubbing the way I lost myself in music. I assaulted the concrete floor with a brush and bleach-filled water so hot it nearly scalded my hands.

I hurled myself down on the concrete to scrub the plumbing under the sink, displacing a daddy longlegs. I was just finishing up when I heard a sound, and there was Daisy, staring at me from the doorway in that disconcerting way she had.

Marjorie Perry's house was a pleasure, with its sleeping porches high above the pine canopy. Daisy only did the kitchen, and then we returned to the main lodge. As we entered, I could hear the clink of silver on china as the campers finished their lunch, and Peeler and Ruby laughing up in the open-screened kitchen. I dreaded bringing my sullen face into their midst, but it was the only face I had.

"Why the long face, sport?" said Pokie. "You ain't got no sugar for me?" I kept my eyes on the floor. I knew if anybody hugged me or said the wrong thing, I would burst into tears.

~~~

The next few days, Daisy and I settled into a rhythm. She would drop me at the Julie Harris Theater in the morning and start at the other end of the camp. Occasionally, when she was running late, I could stretch my work time out long enough to watch Kingo leading a scene study class. I was reminded that somewhere out there an entire world of people spent all their days thinking, living, breathing theater and dance. Daisy's great hope for me was that one day I would "find a decent job," any job, and get out from under her roof. Kingo and Portia Mansfield didn't have jobs, they had a vocation—art—to which they devoted themselves like a religion. Their ideas of what was possible didn't intersect with Daisy's at all. Just being around this other universe began to soothe me. It was as if I was pregnant with the idea that one day, somehow, I might become one of those people of vocation.

After lunch, Daisy would send me out to give the Louis

Horst a once-over while she finished up at the lodge, and this was the magic time. The sight of that simple raised platform at the edge of a grove of aspens, containing nothing but a piano, never failed to excite me. I ran the dust mop along the floor first. Then I went to the piano. I emptied the ashtray that was nearly overflowing onto the keys and then dusted the instrument itself. It was nothing grand, a battered and stained upright with a chipped middle F and faded black keys. But when I sat down and played, in the vastness of the studio, the only motion the wind ruffling the grass in the meadow beyond, it felt as if the notes were flowing out of me, as if I were the instrument, and what Bach had written merely needed my fingers to be released. That mysterious feeling returned, and each day it stayed with me a little longer, so that I began to go about my work without resisting Daisy so much. She noticed the change, and instinctively mistrusted it. "What are you up to, anyhow? You way too quiet."

I began to know which other studios were empty when, and between cleaning sessions, I'd go and play for a bit. The fear that I was going to get caught doing something wrong lifted. I began to riff a bit on what I had learned, imitating what I'd heard the rehearsal pianists do. One day in the Louis Horst I looked up from the keyboard and came out of a fantasy of a classful of nymphs at the barre to see Portia Mansfield standing there, her red bun backlit so that it looked as if her head was on fire. I flushed in confusion. How could I not have heard her approach?

But Portia was unperturbed. "That was interesting, Rita. I had no idea you played." She eyed the mop and bucket I'd left lying on the floor. "Are you going to be joining us this summer?"

My throat was dry as wheat, but I forced myself to get the words out. "I want to," I said.

"Well, the first session has begun, but maybe we can work something out for the second," she said, and then she was gone, so quickly I wondered if she'd been a part of my fantasy.

And I was caught unawares again a few afternoons later. I had just shed my shoes in celebration of finishing cleaning the Silver Spruce Theater and was collapsing on the couch in the lounge with a raspberry Popsicle when I heard Daisy's Jeep idling outside. I was struggling into my shoes when she appeared in the doorway, out of breath.

"Pull your hair together," she said. "Miss Perry wants to talk to you."

My mind raced while I tried to tamp down the braid that had sprung loose, like an antenna listening for the astrals. I still refused to wear a bandanna. "What'd I do?" I said, wiping my forehead on my sleeve.

If she knew, Daisy was in too much of a hurry to answer. She practically ran outside and I followed her, bleach spots on my jeans, cleanser on my shirt. I could smell the cheese soufflé Pokie was baking for dinner, hear the wind ruffling the pine boughs. Someone in the distance was playing chopsticks on a piano that had gone out of tune. As we approached the main lodge, I wondered if they needed the main dining room cleaned for a meeting, although I couldn't understand why Kingo needed to speak to me for that.

"All right now, straighten up and fly right," Daisy said as we hurried through the dining room, where campers on work-study were wiping off the tables. The room smelled of Thousand Island dressing and sour dishrags.

We found Kingo on the phone in her office, looking very Ivy League in her lemon-colored linen slacks and ivory short-sleeved shirt, a beige sweater over her shoulders, the silver pen and earrings a light ethnic touch. The afternoon light spilled in on her frizzy curls that were not exactly brown and not quite gray. She gestured us in and I stepped into the room, profoundly aware that I reeked of Clorox and my shoes were wet. Daisy stayed by the door, holding her hands in front of her as if she was not sure what to do with them. I looked into those clear, confident blue eyes—not hostile, but not cuddly either—and thought, Oh, Christ, why did I ever imagine I could travel with this crowd?

Kingo put down the phone and bounded up, accompanied by her poodle, Andy. "Rita, Portia tells me you know your way around a keyboard."

Now that the moment I'd dreamed of was happening, I couldn't find my voice, but Daisy piped up, "I been giving her lessons since she was a little bitty thing. Rita Ann can play."

"Well, I was wondering if you might play something for me." Kingo came out from behind the desk and headed toward the lounge. She walked fast, accustomed to being followed. We passed by the back door of the kitchen, where a heavy spray nozzle hissed water against a plate and I could see the cooks putting away lunch. In the lounge, the remains of a pine fire perfumed the room. A red, gray, and black Navajo rug from Portia and Kingo's travels in Mexico covered plain, white-painted studs. Against the opposite wall stood another battered piano with tea-colored keys. I took a seat, but my confidence deserted me in the face of Kingo's matter-of-fact directness, and performing Bach in front of this woman who played violin in the Monterey Symphony did not seem like a good idea at all. I would hear no end of it from Daisy if I decided to try out one of my improvisational riffs, though. Kingo pulled a bench up close to the piano, and Andy lay down at her feet, his curly fur gleaming in the soft light spilling in from the kitchen.

Through the kitchen door, I heard America's voice. "Yeah, she fixin to play now."

A pine log popped with trapped moisture, and I flinched. I could feel Daisy, over by the kitchen door, wanting to hit me for being timid.

I put my hands on the keyboard and pushed back the bench, which felt way too low for the keyboard. I thought of Kingo's violin, gleaming on its stand at Cabeen, how the bow shed strands as if undone by the passion of her playing. The piles of music, yellowed and familiar.

And then I closed my eyes and remembered what it felt like to play at Louis Horst, alone and unafraid. "I think this would be right for pliés," I said, and I started the Moonlight Sonata.

As I played, I fell inside the phrases unwinding from the little sour-noted piano. I could see the young dancers at the bar, their little black shoes, pale tights, behinds not quite tucked yet, the music supporting them as they worked on turning their slender hips and knees out. I could see them in first position, their arms not quite rounded enough, fingers a little stiff, the flow of the notes coaxing them to relax as the teacher walks from one to the other, adjusting. The notes, leading each to the other, drew me through, and even though I rushed the ending, which I had not practiced as diligently as the rest, I managed it competently. I rested inside the last chord briefly before I risked a glance at Kingo.

"Well, that'll do just fine," she said. "Portia mentioned you playing in July for the second session. But we're short an accompanist starting Monday. You want to give it a go? In exchange for a couple of classes?"

I nodded, transfixed, then glanced at Daisy. She would be wanting me to earn something.

Kingo looked at me, all business, as though she was expecting a more definite response. Finally I coughed out, "Yes."

She smiled her no-nonsense smile and turned to Daisy. "Can you have her here by eight?"

Daisy didn't look at all surprised. "Of course," she said. "But we need to go now and finish up the Julie Harris."

~~~

Sunday night I wallowed around in fitful dreams about what I would play for the dance class. But I must have slept eventually, because I woke to the sound of the stove lid sliding into place and the familiar smell of burning kindling. Daisy was strangely quiet at breakfast. She seemed off her stride, as if the habit of flogging me toward success was the only mode she knew, and its actual arrival had thrown her. I was just feeding Jiggs when Daisy came down the stairs, saying, "Don't be messing with that dog. I've got to get you to camp early. I'm going into work for Mrs. Leckenby today."

The Leckenbys, who owned the *Pilot*, were Daisy's staple winter and summer job, and I knew she had taken good care that they not feel shortchanged. By seven, when she pulled over by PM's front gate to drop me off, she already looked tired, and her bandanna was askew. I felt sorry for her, but I also couldn't wait for her to be gone. She kept her gaze on the rearview mirror and her foot on the clutch as she said, "Be here at five. Don't make me have to come in and look for you." And then, as if in an afterthought, she reached behind her and handed me a plastic bag. "You'll be needing these. All right, out you go."

The wheels of the Jeep sprayed gravel as she spun off and I jumped back. I opened the bag and discovered a faded leotard and tights that Daisy must have rescued from the trash.

A group of campers was already in the corral. They were

the littlest girls, on horses so big they must have had to stand next to the fence to mount. I wondered if this was the group I would be playing for this afternoon. From the center of the corral, Shannon, the equestrianship director for as far back as I could remember, called out to one tiny camper who sat clutching her saddle horn. Her hat had bounced off and hung from its string down her back, and her head flopped as the horse trotted around the ring. "Post, sweetheart," Shannon called to her. "Use your legs." As she began to rise and fall with the horse's rhythm, a smile of wonder broke through her consternation and she straightened up in the saddle. Same spine as she'd use in a relevé this afternoon, I thought to myself.

I continued on past the stables, delighted at the prospect of a day without interference from Daisy. The class I was to take started at one. As I walked up the hill to the main lodge, I smelled something delicious baking. From the Silver Spruce, I could hear lunging dancers and a teacher yelling, "ONE two three, ONE two three." I resolved to get a cookie from America's cooling rack and then go watch them.

When I walked into the kitchen, I nearly backed out from the heat. The soup pots were going on the range, their lids vibrating. Pokie was about to tuck a fistful of stuffing into a twenty-pound turkey spread out on a wooden board. Blisters of sweat gleamed on her forehead, and she squinted at me through the smoke from a Camel hanging from her lips. It never ceased to amaze me that she could maneuver all manner of food from pot to counter without once losing track of the ashes.

America, already top-heavy, was bent over a rolling pin and a disk of dough the color of vanilla ice cream. When she saw me, she burst into a delighted smile. "All right y'all. Here she come, tippin like a Maltese kitten in the mornin dew."

I allowed myself to be enveloped in her fragrant arms. Pokie said, "We heard you play for Kingo. When did you learn to play that good?"

"I'm not that good," I said, secretly hoping my modesty would lead to another compliment, but America said, "Don't nobody start out good. Aggie done started goofing around at a square dance when she was here at camp. One thing lead to another, she ended up choreographing *Rodeo,* and eventually *Oklahoma.*"

"Aggie?" I said.

"Agnes de Mille," she said, turning back to her dough. "She was before your time. Crazy 'bout horses and dance, just like you."

Pokie was trussing the turkey. "America remembers every single camper ever come through here, you know."

"Toni—Antoinette Perry—loved theater from the beginning, but she didn't expect to end up a rich widow with the means to do something about that love. You trust in the Lord, honey, you'll be somebody."

I had been feeling fine up till then, but now I felt as if I'd been caught in a chinook—warm and balmy one minute, icy the next. America, who had served Kingo in New York, in Steamboat, in Carmel by the Sea, who remembered the name of practically every single camper ever to come through Perry Mansfield, was still working in the camp kitchen, however many decades later. I turned and nearly ran back out into the smell of pine sap in the cool air, but I could still hear America's robust, tuneless voice singing, "His eye is on the sparrow."

I was grateful that there were only six students in the class I had come to play for, none more than eight years old, all of them new to ballet. The girl who'd been having trouble in the corral that morning was there, with a sunburn on her nose. The instructor was Missy, a beautiful honey blond with extraordinarily mangled feet who'd been coming to the camp for a long time. "I'm so glad you're going to help out," she said. She surveyed the class, spine elongated, head straight, and each dancer, taking her place at the barre, unconsciously imitated the stance.

The class was still working on the basics of the five positions, so there was only very intermittent need for music as Missy went from one student to the next, correcting, awakening, encouraging each dancer to relax her shoulders, extend a leg. Almost before I knew it, the session was over, and I hurried to the lavatory to change for my own.

The tights had lost their elasticity, and as I pulled them on they gave off the odor of mildew. Daisy had wadded them up in one of the plastic bags we washed and reused, so they also smelled of the soap she made from bacon grease and lye. The heels and toes of the tights had been cut off, leaving only a stirrup under the arch to keep the dancer from slipping on the wooden floor. A runner up the heel needed mending. The muscles of my thighs were tight in the broad place where Daisy used to smack me with the willow switch she made me cut with my pocketknife. As I stretched the nearly sheer cloth across them, I remembered Daisy had told me once how Mr. Anderson had flinched when she tried to touch the scars on his back. I had heard somewhere that memories got lodged in muscle. Would I actually be able to dance? I had spent so much of my life cowering. I pulled up the waistband of the tights, wondering if the same muscles that had flinched from a hiding could now lift aloft in a graceful arabesque. Would I stumble, remembering that iron-eyed nun and her look of revulsion as I danced that night at Mount Saint Scholastica, riding the music like a horse in the night?

I put on the leotard and came out of the stall. I stood before the mirror, contemplating the undancer-like roundness of my haunches. Daisy had said they bred slaves like draft horses, to bear heavy loads. The neckline framed my chest gently, but the tights bagged in the knees, and the seat of the leotard drooped like a loaded diaper. Would I catch my reflection in the studio mirror and cringe at the spectacle of myself trying to prance on my toes, like a pixie on a lily pad?

A redheaded camper came in. She got a bar of soap out of her cubicle, glanced at me, and smiled. I looked at her leotard and realized it was rattier than mine. I remembered that many times campers washed their dance clothes in bleach just so they'd look worn.

"Hello," she said, open as an unmowed field.

"Hi, I'm going to a modern class," I blurted out, as if that was the most natural thing in the world, then flushed with embarrassment.

But the redhead didn't seem to think anything of it. She commenced scrubbing her blotchy face. "Who with?" she asked, constricting her mouth.

Paralysis gripped me. I didn't know. "It's at the Louis Horst, at three o'clock." I said. That was all I remembered.

The redhead toweled her face, then pulled her hair back into a bun so tight it looked like it hurt. "Harriette Ann Gray is taking over that class today, I think."

This worried me a little. Harriette Ann Gray had her own company and was the head of the entire dance program at Perry Mansfield.

"Make sure to show up early," she said. "Harriette Ann expects her dancers to have warmed up for fifteen minutes before she walks into the studio."

When I walked down the steps into the wide studio with its sides open to the surrounding forest, its barre, its tottery piano, and its gleaming floorboards, I nearly laughed out loud. How many summers had I stood outside watching classes while Daisy mopped the floor that dust almost instantly reclaimed? Outside a clump of slender aspen saplings reached for the clouds, their leaves trembling in the hot breeze. The tin roof still held the heat from high noon, and I was thirsty. Looking down at my stomach, I wished I'd had the self-restraint not to eat so much of Daisy's fried chicken when Grace and the uncles were visiting.

Twelve dancers strode in, coiled as cats, their gleaming hair pinned back into little buns at the napes of their necks. Only a minute before, in the Green Room, they had been giggling teenagers, but the studio itself seemed to transform them, lengthening their strides and taking them inside themselves. They dumped their enormous bags by the wall, plucked delicately pointed toes out of their sandals, and carefully began to stretch and flex each body part: toes, calves, arms, back, neck, throat, fingers, and chest. Although I didn't know their names, I recognized many of them from previous summers at Perry Mansfield. They were at least a year older than I, with real bosoms and shockingly limber legs. Everything about them was lithe and smooth except their feet, which were all knots and tape, calluses and bunions. I realized that the soles of their feet were as hard and calloused as my hands, which had been toughened from handling shovels and wire stretchers and scrub brushes and rags. Of course—they had been dancing the entire winter. How supple they were, unraveling their spines from the hinge of their pelvises, flinging their arms and heads down nearly to the boards. There was an edginess to their abandon, however, as if they were strutting their stuff for one another, a whiff of competition on the wind. And they did not smell of mildew or lye.

I pulled up the sagging seat of my leotard and edged to the corner of the studio, trying to lose myself in the safety of the pine studs. Outside a woodpecker drilled pine bark for grubs. I arranged my limbs into second position. *Plié, relevé, plié. Plié,* reach left, arms center, lean right. I stretched long, from the tips of my fingers to the end of my smallest toe, and the stretch was solid.

Then suddenly all sound and movement ceased, and Harriette Ann Gray materialized on the stage above the studio, as if she were made of the cigarette smoke that wreathed her. She seemed to hover in the very air, the vast space behind her framing her fierceness with nothing. Calmly she stood there, her body sheathed in a leotard and tights the color of charcoal. She picked up an ashtray, took one last loving drag, and, while the smoke was drifting out of her mouth, ground the butt into the glass and surveyed the studio.

The dancers stood motionless, hardly breathing, as she looked us over, her eyes coming to rest on each of us in turn. The space practically vibrated with anticipation. Instinctively, I stood taller, remembering Keoko saying, "Reach beyond any height you have ever tried for, create light between the joints, air inside the spine, openness, space, room to move." It struck me as odd that both nuns and dancers chose uniforms of black, that they reached for something higher with every breath they drew, that the vocation they had chosen required a subjugation that was absolute. But while the nuns draped the body and considered it borrowed, a vessel for the soul to roost in only until they could shed it for a seat at the right hand of God, the mortification of the flesh that was dance was attained by exposing the body and inhabiting it with vigor and discipline. I took a deep breath for what felt like the first time in months.

"So, how are we this afternoon?" Harriette asked. "Hm?" She had an irony about her that made you want to lean forward, to be in on the joke. Such graceful hands, as if her fingers were winged. She drew her arm across her chest and rested it there, as if she were planning what to do next. A gray streak at her temple looked like a bolt from Zeus. She continued her inspection, greeting some, merely smiling at others, until her eyes came to rest on me, pasted to the lumber in the very back of the studio. This is it, I thought. How many times had I hidden under Daisy's oak table, practicing stillness the way a fawn does, until danger has passed? Stillness would not save me now.

"Are you joining us?" I couldn't speak, so I nodded, terrified.

"Well, come on then," she said, pointing to the very front of the class. All eyes turned to me, and I wondered if my fear was visible, drifting off me like the snow spume from the top of Storm Mountain. Was my sudden sweat making the smell of mildew and lye bloom in the studio? Still, there was the absolutely thrilling possibility that I could do what Harriette was asking. I glided to the corner of the second line of dancers, next to a blondish girl with ripped leg warmers wound around her ankles like something Daisy would rig to contain her hernia. Her leotard had holes cut in the front so that the cubes of muscle in her abdomen could reveal themselves as if by accident. She scooted closer to the center to make space for me and flicked me a smile so kind it made my eyes well up. Then I was among them, flexing my arm across the front of my body like a crane stretching a wing.

Harriette raised her eyebrow and surveyed us. "So, I see we all look pretty." Instantly we drooped, crestfallen and sheepish. "Is that all we do as dancers? Point our toes nicely and look lovely in pink?" A ripple of embarrassed shuffling traveled the length of the class. "I am not interested in 'pretty,'" said Harriette Ann. The girls glanced at each other sideways, the air electric; nobody came here without dreaming of becoming a beautiful dancer. Harriette continued, "Let's consider how we inhabit space. Dance is not only meant to be pleasing. It's also about struggle." That glance passed over each of us again, not coldly, but with a terrifying ferocity. When it came to rest on me, the skin on my neck rippled with the current. And then she turned her head sharply to the right, just as an invisible hook grabbed her left arm and pulled her so hard she fell to the floor, but without sound, as a leaf might fall. Then in one smooth flow she rose again, as if another invisible force had picked her up and placed her upright again. "We have a responsibility to include all the vectors of human experience in the dance—not just the fluffy ones."

There was a ripple of nervous laughter, then the studio went silent as we took in this idea, relaxed into it. Harriette Ann continued, "I am interested in what happens before you move. What happens at the core? What are you thinking about? As a matter of fact, are you thinking? Did any of you go and see the Dance of the Minoans last year?"

Everyone nodded, including me. I had gone, and it had rocked me. In fact, I had thought about that performance all year. An enormous form of a bull, the dancers leaping over it as if they were spirits possessed.

"You don't believe, do you, that the ancestors of the Greeks came to worship at the altar of Cute?" Harriette Ann slumped over again, then, as if that hidden force had slapped her, leaped sideways, arms and legs jagged and sharp.

"Give me an extension that starts with a need. Dance is just like acting. You never just say a line. You are motivated to accomplish something and then the lines you speak are a *result* of that. Give me an extension that starts with hunger, anger, grief, and carries you all the way across the studio."

She sent us to the corners of the room, admonishing us, "Think, people. Make this personal."

Briefly it occurred to me that I could just leave, run away. But before I had time to move, the other girls were lining up, three abreast, and I was among them.

The first group was surprisingly strong, but decidedly ungraceful. Harriette Ann nodded. "That's it. Even your skin has to be engaged in the work." She surged around the outside of the studio, the afternoon sun flickering on her face, the edges of her feet brushing the hardwood floor as if they were searching for purchase. She seemed almost angry, or malevolent, like a stalker. At last she came to rest against the barre, and the students swiveled as one to face her. Chest high, head back, and legs long, she said, "When you move, let's see those extensions lead you across the room because they *have to*. Remember, dance is drama."

When the two other girls in my line stepped forward, astonishingly, I came with them, and when they leaped, I leaped. I felt a surge of exhilaration until a sharp handclap interrupted my reverie.

"Dancer in the brown. Stop watching yourself."

I looked up, startled to find Harriette Ann looking at me. How could she know?

"What's your name?"

"Rita."

"Well, Rita, stop performing. You have to do the process to get to the performance. In fact, you've stopped breathing altogether and gone straight into the concept of movement rather than being in the impulse itself. Do you understand?" She came toward my corner, peered into me. It took everything I had to stay put.

Harriette Ann grabbed my waist and nearly catapulted me sideways, so that I was forced to extend my leg to recover. I saw her register the tears that had begun to well up in my eyes, but she didn't back off. "It's good if it surprises you." She nodded her head just slightly as if to reassure me, then grasped my arm. Her hand was cool, hollow as a bird's bone.

"Relax into it. Trust it. Let it catch and carry you, lead you. Come on."

I wondered if this was what it meant to be a novitiate.

~~~

When the first session ended, I was no longer needed as an accompanist but Kingo offered to let me continue in the second session as a full-time camper. To my astonishment, Daisy acquiesced. She dropped me off at seven thirty in the morning, when clouds still lay in the meadow. I had ballet at eight, modern at nine thirty, riding at eleven thirty. I was secretly delighted at how much my body had changed since June, under the daily discipline of classes. My abdomen was always taut now, and a core awakening had opened my torso and lengthened my spine. Acting class in the afternoon challenged me in ways I hadn't expected, bringing to the surface feelings I had barely been aware of holding at bay.

In the final two weeks I was cast as the Negro cook in a scene from the stage version of Carson McCullers's *The Member of the Wedding*, a role I greeted with intense confusion. On the one hand, I recognized there were precious few parts for a black girl from a redneck background who spoke German and loved Geronimo, and Carson McCullers was no slouch of a writer. On the other hand, when I'd sat in the dressing room of the Julie Harris, I hadn't pictured myself learning to say "Listen to 'em," "dis" and "dat," and "baffroom," in the very Southern accent Daisy loathed. I had imagined myself wearing brocade gowns, not old lady stockings rolled down at the knee. I didn't dare tell Daisy that I'd landed the role, because I knew she'd be furious.

Of course, she found out.

"They had you playing a Negro maid?"

"You heard wrong. She wasn't a maid at all," I replied. I didn't tell her she was a cook.

By the middle of August, two months of trucks, hooves,

and feet had left a veneer of dust on the camp as fine as Daisy's face powder, coating the leaves of the skunk cabbage and Indian paintbrush so heavily that their tops drooped down into the dirt. And then all at once, all over camp, people started pulling down trunks and packing. The hay bales were set outside the main lodge to be picked up, and the fireplace swept clean. And then everyone left in a rush, and camp was deathly still, but for the occasional pounding of a hammer as another studio or cabin was boarded up.

Soon fall closed in, like the steel clamps of a coyote trap. Dusk was no longer a cool refuge but an interlude of minutes before cold night set in. In the morning, the dew froze on the hay, so that the blond chaff seemed to have been sprinkled with diamond dust. Frost extended its reach across the pasture, turning new shoots of mowed grass brown and burned, seasoning the land for the layers of ice that would soon settle in, sixty inches deep. Every dawn, a little more green drained from the aspens until their leaves were a yolky gold, while the maple leaves turned to five spiky points of rust. Clouds nested on a clump of blue spruce above Bear Creek Pass.

The summer residents had sped off to places where important things happened—Chicago, Los Angeles, New York, Denver. To them, Steamboat was just a girl on the side, a bit of holiday fun. The slow-driving locals reclaimed the roads. Time to think of the final harvest, school clothes, hunting season. Daisy began to feel her arthritis again, as the temperature dropped and stayed low. I struggled to hold on to the images of dance, the faces of my friends. Daisy filled the silences, but when I tuned her out, she complained. I wanted to be grateful for the magnificent summer she'd given me, but after the surge of possibility I'd felt at Perry Mansfield, I could hardly stand to listen to her. She, on the other hand, wasn't sorry to see "that bunch" get on down the road. Six weeks of tutus was "aplenty." Now it was time to get down to business. Get out the bluing oil for the rifle. Store the root vegetables in the cellar.

Perry Mansfield had opened gates in me that I couldn't shut. Now I knew that Sally was going to study cello in Los Angeles, where her father played in the symphony, while Sidney was going to work with an acting coach in Chicago. I had met their pretty, fragrant mothers who made lunch, and their competent, prosperous fathers who made the mortgage, and I was filled with a more precise rage. "Daisy, you've got to promise me I can go back to camp next summer. Will you please promise me that?" Daisy had already forgotten them. The most she would give me was, "If the good Lord's willin and the creek don't rise."

Later, she sat in the big red chair in front of the washing machine with a whetstone, sharpening the blade of the ax, slowly, rhythmically, the rasp in rhythm with the flash of steel in the light from Gama's coal oil lantern. The season of slaughter was upon us.

~~~

"**What I really** need is a good hog's head," said Daisy, "but she probably ain't going to give me that." We had driven out toward Oak Creek, to the Ames ranch at the foot of Rabbit Ears Pass. An immaculately weeded potato field stretched the length of a gridiron in front of the house. The plants looked vibrant and well tended, although the house itself needed painting. These folks were wealthy, but country wealth meant passing the ranch from one generation to the next and hardly ever buying anything flashier than a bright green John Deere tractor. Their place was as littered as ours, but on a much grander scale. Every car, tractor, and combine this family had ever owned seemed to be stashed somewhere on the property. Out back, a wide pasture spread back to the mountain that rose in the distance, as if the ranch had been tucked into the valley for safekeeping.

Daisy stopped short of the barn, surveying the expanse of mud in front of it. "You'd think they'd want to put down a load of gravel wouldn't you?"

A dog barked sharply, announcing our arrival, and the door to the house opened. A tall, stout woman walked out on the covered porch and waved to us. Even across the yard, her housedress looked faded, and she wore men's shoes caked with mud. She'd probably been out in the field early, digging potatoes. My friends would have called Enid Ames fat, but her weight didn't slow her down.

"You stay here and I'll be back in a little bit," Daisy said, shutting the door of the Jeep.

In the distance, a truck downshifted to descend the pass, its engine groaning to slow the load. The fitful wind made the rusty hinges on the hayloft doors creak as they banged aimlessly open and shut under the eaves of the barn. "Hi," I said out the window to the mistrustful border collie, who had stayed behind to watch me. She raised her head from her paws, her bright eyes neither unfriendly nor inviting.

The mumbling hens were beginning to stick their heads out the door of the coop, searching for their morning corn. They kept settling down on the ground hoping for a nice warm dust bath, but the dew had muddied the powdery soil. The loamy smell of freshly turned earth filled my nose. Near the fence just beyond the Jeep, a faded scarecrow, his job complete for the season, seemed on the verge of collapse.

The book I'd brought along to read reproached me. It was the biggest book I'd ever tried, *The Rise and Fall of the Third Reich*. A counselor at Perry Mansfield had suggested it, but the words were hard for me, and there were no pictures. I still wasn't sure what Jews were exactly, but I found it hard to imagine how somebody could get away with killing six million white people when Ernest had gotten carted off to the pen for swiping a calf.

Thou shalt not kill. When did it matter? The Ute arrowheads we were always finding showed that they had been successfully eliminated by people like ourselves.

I knew I was not so innocent, either. I'd let my little duck, Donald, die when I was seven, forgetting him in the shed with no food or water. Daisy took me out there and made me look at his limp body. She hadn't mentioned him or lifted a finger to save him because, damn it to hell, I needed to be taught a lesson. Animals were livestock. If I was going to turn them into pets, I could damn well take care of them. "You can quit your bawlin," she'd said. "Little late for that now."

A screen door slammed and the dog leaped up, barking. "Say, Reeter," Daisy called from the side of the house. "Come here and help us with this butchering. Enid's done gone and jerked her thumb a-loose."

I could feel my face flush. Daisy had mentioned picking up a hog's head. She hadn't said a word about butchering.

"Enid, call your dog off. Reeter scared to get out of the Jeep. She just settin up looking like a bay steer."

"Blackie, get over here." Enid's high girlish voice seemed at odds with her bulk. The dog slunk toward her.

"Enid's giving us this whole lamb. Ain't that nice?" Enid stepped forward, her thumb wrapped in rags, leading a lamb by a rope tied around its neck. Would its head be boiling on our stove in pickling spices come nightfall? Would I have to help singe off its wool with brown paper, like we did pig's bristles? I knew if I begged them not to kill it, Daisy would peel my hide when we got home. But if I didn't speak up for this little lamb, how could I expect any shepherd to protect me?

"Come on, girl, help me with the gate," said Enid. Daisy didn't like being called "girl," and even in her hurry to get the carcass home, cut up, packaged, and transported to the rented meat locker, I knew it wouldn't be forgotten. The lamb struggled against the rope, its little hooves so shiny it looked as if they had been polished. Its bleating brought the collie to her feet, although she stayed low to the ground as though trying to hide.

The butchering shed could have been ours—the rifle rack by the door, the plastic on the floor, a slab of plywood nailed onto sawhorses, a chain with a hook hanging from the ceiling so that the blood and guts would spill out of the body cavity smoothly. The Ameses probably aged their deer here in the fall just as we did.

On Sundays Enid wore a hat and a little lipstick to Mass. I wondered what she thought when the priest talked about the Precious Lamb of God. What did it turn on, this random business of death?

"Can I feed it?" I asked. "Can it have some more milk before it dies?"

"Naw," Enid said. "You want the bladder and guts empty as you can get 'em when you slaughter it. Pee and manure'll taint the meat if the gut bursts." Enid had curiously narrow shoulders, and her fat arms looked weak, but she easily lifted the struggling lamb onto the plywood platform. Even with her bandaged thumb, she was able to tie each front hoof to a nail.

"Did your beans make?" Daisy asked companionably, as if the two women were a pair of hens roosting side by side. Enid shook her head as she shielded the lamb's eyes. Did it smell the blood that had been spilled here before?

"That's a shame. Frost took mine too. You go to all that work."

Outside the collie barked frantically.

"Where's the whetstone?" Daisy asked, testing the knife with her thumb.

"Over yonder," Enid said, pointing with her chin to a drawer under the cabinet. The lamb, unaccustomed to being handled roughly, urinated, but Enid deftly moved out of the way, so that it didn't hit her shoes.

"Daisy," she said, "the meat'll turn if it wrassles around here too much longer. That blade is plenty sharp."

"All right, now Reeter," Daisy said. "Help me hold it so its feet don't kick us in the face."

The sun, higher now, shone through the window onto the lamb's lanolin-soft wool. It lifted its head and looked back at me as Enid stabbed it in the throat. "Oh, God, how could you do that?" I wanted to cry out.

The lamb died surprisingly fast. The bucket under its throat filled rapidly with blood. Finally it fell. It seemed to sigh once and then was still.

Daisy could see by the expression on my face that I was thinking about being "troublesome," and she smiled me a cold warning. I said nothing.

Deftly Enid sliced open the belly, taking care not to puncture the silvery casing that contained the guts. "The heart and lungs'll make a nice stew," she said. "What the heck, I'll throw in the liver for you all too."

I could no longer contain myself. "How did you know it didn't have a soul? How do you know it's not the Lamb of God?"

Daisy stepped between us. "Rita. Sugar. You need to hush up now. Enid's gon think you a mental, sweetheart." Enid looked up at me as if she were seeing me for the first time. "It is one of the good Lord's lambs sent right down here for us to eat," she said.

10 ~ Whiteman

Daisy in her sixties

~~~~~ **September came,** and I still had no idea where I was spending my sophomore year. Daisy was moody all the time, and increasingly forgetful. She lost her keys at least once a day and chattered to herself constantly, whether I was in the room or not. One afternoon after going to the co-op to pick up poultry feed, we stopped in Boggs Hardware. Mr. Boggs watched as Daisy picked up a claw hammer and put it back, then

picked up a crowbar and replaced it. "Daisy, can I help you find something?" he asked.

"Done got old now," Daisy replied, as if that answered the question. I walked over to the nail bin, trying to squelch my embarrassment and bracing for the recital I knew was coming. "My knees hurt me so bad. And I got a pain come in my hip." When we finally made it to the cash register with a lone box of staples and Daisy sifted through her coin purse for the right combination of change, Mr. Boggs brought the entire transaction to a halt by tucking a small calendar into the bag.

"Ain't that just the prettiest thing you ever saw?" Daisy rhapsodized.

"Daisy, can we just go, for Christ's sake," I said, with more anger than I meant to show. They both looked at me then, startled at the impertinence.

On the way to the post office to check our mailbox, I noticed cute red and white cheerleader skirts in the window of the Dorothy Shop.

"Daisy," I said, "if I'm going back to Steamboat High, I'd like to try out for the cheerleading squad." I knew she'd never go for that, but I had to ease her into the subject somehow.

To my surprise, she answered immediately. "You're not going back to Steamboat. You're going to Whiteman, starting next week."

I nearly fell out of the Jeep. Every single time I thought I knew where Daisy was, she banked off some plane I hadn't even known existed. This time, no news could have made me happier. Like all-girl Perry Mansfield, Whiteman, a coed college preparatory school with an enrollment of barely thirty, attracted students from all across the country. Its progressive vision included time abroad for students to expand their cultural horizons. I knew better than to hope that I'd get to live on campus and finally learn to ride and ski properly, but I was thrilled nevertheless.

Monday morning, when Daisy drove the Jeep up Hot Springs Road, I assumed she was going to drop me off for registration and take off for the Leckenby house as usual. Instead, she got out of the Jeep along with me. When she went around back and grabbed her mop and bucket, I got it. She was going to work at Whiteman, presumably in exchange for my tuition.

"I'll pick you up after classes," Daisy said. I didn't even look at her. I couldn't trust myself to speak. I was aware of what a sacrifice she had made to send me here, and how ungrateful I was being, but I already knew that in her own way, Daisy was going to make me pay.

Whiteman School had the woodsy, rough texture of a ranch. It consisted of a main lodge where all classes and meals were held, a stable, and surrounding dormitories that resembled bunkhouses. An old wagon adorned the driveway, and the spacious pasture out back was lush and well tended. But I was not fooled by the knotty pine. There was nothing hokey about Whiteman's reputation for tough academic standards and terrific teachers.

Now that the school was mine, the place took on an entirely different cast. To be sure, I was intimidated. When we gathered for our first group meeting in the bare-bones main room that served as study hall, dining area, and classroom, I could hardly hear a word that Lowell Whiteman said, I was so dazzled by the poise of my new peers. Not only was I one of the youngest students, I was also the poorest, one of only two day students, and the only black girl. I knew that my classmates' turtlenecks and ski sweaters cost more than Daisy earned in a month. I had met wealthy kids at Mount Saint Scholastica, but there was something different about the Whiteman students. They looked like they had stepped out of a Beach Boys album cover, the boys with their surfer hair hanging across their eyes, and the girls with their straight, silky bobs that swung just so. I was sure not one of them had ever mucked out a henhouse before racing for the school bus.

I was so cow-eyed with shyness that I could barely speak when I was introduced to three students from the same big family in Aspen. They had eyelashes long as awnings, big dimples, and identical generous, open smiles. Completely at ease, they greeted me with such surprising openness and spontaneity that I could only stare at the floor in confusion. I had learned that beauty was its own currency and never expected pretty people to be anything but ruthless.

And after my stint at all-girl Mount Saint Scholastica, I was both delighted and confounded at being in the company of boys again. I was well aware that Daisy believed the women of Steamboat Springs found her liking for white men scandalous. And I knew that behind all the sudden displacements to Catholic school and now to Whiteman lay Daisy's fear that I would get a reputation for being sexual, threatening her already dicey status in the community.

Mr. Whiteman introduced the teachers, most of whom had a maverick gleam in their eyes. Nobody came to a school this far out of the way if teaching was just a job. Mr. Kakla, who was renovating a barn to live in, looked like he had stolen his beard from a grizzly. Madame Bair was an exuberant Frenchwoman with an endearing, eccentric accent. Mrs. Garland, who would be my German teacher, seemed to have chipped off her own piece of the sun. John Whittum, a patrician Easterner who had just rented the cabin my grandparents used to live in, would be teaching history.

From the very beginning, not one of these teachers gave me

the sense that they had reservations about me, racially or otherwise. And I in turn was a shameless teacher's pet, hungry for parenting and some buffer from my daunting peers. There were only three of us in my German class, and I already had picked up enough of the language that I had a feel for the way the syntax worked, the verb stuck on at the end of a sentence like a caboose. Much as I'd struggled with algebra at Mount Saint Scholastica, it turned out that I loved geometry and had a knack for it. Each formula was married to its own little picture, I discovered, and that allowed me to grasp it better. I had never thought I was all that bright, but the classes were so small and the teachers seemed so excited about their material that I got all the attention I needed. For a while, I forgot to be afraid.

Even Mr. Whittum, the most formal of the lot, held me absolutely rapt in his ancient history class. On our way home one evening, I recounted to Daisy how Hannibal had brought a platoon of elephants across the Italian Alps to wage the second Punic war. She was dumbfounded. "I wonder what all them poor ellyfins et," she mused. I knew a little about war—sheepherders hated ranchers, and ranchers could not abide "mountain maggots." But in the *Iliad* I discovered that the Greeks had created a whole platoon of gods and goddesses as their own personal special forces. They made up religious myths to support their needs and wishes, just as I suspected we had, with our definitions of the divine, and our nuns and priests whose role it was to intercede between laypeople and God. What was the difference between the Greeks making up their gods and the Roman fathers coming up with theirs? Who decided there was a purgatory and a hell, a Hades and an Olympus? Why did there have to be a blessed trinity? There could just as easily have been a blessed four or two or twenty-five. Who decided that Jesus' drab, sad mother needed to be a virgin? Or if she had to be a virgin, why couldn't she shoot straight like Artemis, or command the essence of beauty like Aphrodite? Why did her womanly power have to be subsumed, cloaked, made mute? Why should the natural seasons of rut, breeding, and birth be labeled savage, filthy, degraded, abominable in the eyes of the Lord, when He—if it was a He—had set up the entire operation to begin with?

I couldn't imagine any of Zeus's children wearing a bloody crown of thorns and asking, "My God, why hast thou forsaken me?" without some other god arranging an intervention. In the *Iliad,* some god or other was always meddling, in a series of episodes that seemed to be just stories people made up. If that was the case, what was true and binding? What, in fact, was fact? Was a Cyclops any more fanciful than the Holy Ghost or "a band of angels comin' after me"?

John Whittum's class made me reexamine all kinds of things I thought I knew, and not only about school. I was intrigued by his commitment to teaching, and by his decision to live way out by us in Strawberry Park, with its frigid temperatures and country pace that bored me beyond description. With his Ivy League education, he could have gone anywhere he wanted. I noticed that he was respectful of Daisy and took an interest in her peculiar history, and in the history of the area. Daisy thought of an education as leverage for power and money. But John Whittum, as far as I could see, appeared to experience it outside those entrapments, as a good in its own right.

Whiteman awakened my intellect the way Perry Mansfield had engaged me as an artist. I could look up at any time and see other students asking deep questions too. Absorbed as I was, I forgot how poor I was, how white they were. There just wasn't time to correctly conjugate twenty irregular German verbs through present, past, future, present perfect, past perfect, and future perfect and malinger on my sorry lot too. After diving into my books in study hall, and then another three hours at home as well, plus chores, I just wanted to sleep.

With the intense studying and my day-student status, I didn't have much time for extracurricular activities, but I did try soccer, and even managed to buy a pair of downhill skis on credit, though I wasn't sure how I was going to learn to use them come winter. Polo was another matter. I had assumed that, as a pretty fearless rider on a good horse, I could stand up to the best of them, despite my ignorance of the game. But my bad right eye gave me trouble with depth perception, and I had been hit so many times as a kid that I was actually afraid of the ball. With the horses galloping first in one direction and then the other, and the ball always between their legs, it seemed an unnecessarily dangerous game. When I took a wild swing at the ball one afternoon and connected instead with the right front hoof of my mare, I decided that the sport was not for me.

But among a sophomore class of barely ten students, I made two good friends: a blond boy from Denver named Tom, and Jill, a girl from California. We whispered together in class, sat in study hall together, and smoked together. When I stood outside the office at the end of first quarter and discovered I'd made the headmaster's list, they rejoiced with me. As we headed into the Christmas holiday, I was thinking I might be able to succeed at the Whiteman School after all.

Daisy, however, seemed to be unraveling by the day. The long hours working at Whiteman and the cold weather seemed to be especially hard on her in this her sixty-fourth year. I didn't allow myself to think what it must be like for her to watch me get the kind of education she would have killed for. But it was obvious she found my demands for material things and rides home late at night after study hall wearing. It was not my least selfish hour, but I was a teenager, and neither of us knew that that is how teenagers are. All my energy went to avoiding her, getting around her, and masking my happiness from her.

Underneath it all, I still longed for her to love me, whether I loved her or not. And I think she was confused to find herself not liking me much, even as I was accomplishing the very thing at which my mother and sisters had failed. We reminded me of the way aspens grow, in dense family stands, saplings and their scarred elders trapped or nourished in the same cool soil. Sometimes a sick old tree will collapse against the saplings that stand close to it, and the saplings in turn grow malformed, spindly and frail.

Daisy was neither well nor happy. Even working as hard as she did, she was failing to make ends meet, and every month or so she would snip off and sell another acre, or two or five. It felt as if every morning I woke to see fences squeezing closer to our front door. One evening when I walked Jiggs up the road, I saw new people moving into our upper cabins. I was about to go over and confront these trespassers when some instinct told me to hold off. When I got home I said carefully, "I see we have new neighbors."

"I reckon we do."

"They gonna be there long?"

Daisy cast me that steely look that warned me not to proceed. "They might be."

Helplessly I watched as she lost both the land and her sanity, bit by bit. I wasn't sure she would hold on to some small parcel for me even if she could. I suspected she couldn't stand the idea of me succeeding on the very spot where she had failed.

Watching the civil rights struggle every night on television and mopping floors by day aggravated old pockets of rage and grief that seemed to be dragging her down further each day. She repeated the same stories over and over now wherever we went, to anyone who would listen, at the drugstore, the five-and-dime, the lumber mill. How she had nearly been killed for not stepping off the sidewalk when a white woman approached. How the Koo Klukkers had run her family off their land. How many properties lost, how many relatives dead.

One cold January morning, the inevitable happened. Ten o'clock came and went and we sat in class waiting for the geometry teacher, whose car had gotten stuck in a ditch and was going to be late. Bored and unsupervised, the class began to get rowdy. We were only eight or so, seated around a long rectangular table in the small classroom. Looking around, I noticed piles of dust and dirty windows, which made me worry that Daisy might lose her job. In the back of the room, the two class troublemakers, Dale and Ron, had drawn the goofy hypotenuse of a cock and balls for geometric study.

Then they commenced to argue about the best powder runs in Colorado. "Storm Mountain's best is nowhere near as dry as Snowmass," said Tom, who seemed to have skied every mountain on the planet.

The argument took on heat as the class became more and more restive. But I was no longer listening, because an old, nearly disfigured black woman had slouched down the hall, dragging her mop behind her as if she was carrying the cross up Golgotha. She came to a stop just outside our classroom, where she gazed out the window, mumbling to herself under her breath. Shame and fear, pity and sorrow all collided in me. Should I get up and try to get her to go back the way she came?

Although I didn't actively deny my connection with Daisy, I hadn't gone out of my way to announce that she was my aunt either. Most of the time, I tried simply to focus on being a student and didn't interact with her much, and I wasn't certain that my classmates had made the connection. Today would not be a good day for the two of us to become the center of attention, especially as it looked like she was about to crumple in front of us all.

Daisy swayed a little as though she had a case of snow hypnosis, seemingly unaware of our presence. With her toes turned in, she stared transfixed at a snowbank, her head tilted to one side. Her whole body sagged, like an old shed with a big load of snow. Her breasts hung unsupported in the hand-me-down bra, her arms were slack, her hernia protruded from her stomach large as a melon. I glanced around to see if the other students had noticed. Had they seen how her jeans were safety-pinned at the side? How she absently picked the pills of wool off her sweater at nipple level? Had they noticed the useless finger she'd lost in that wheel spoke, the mushroom-colored flap sewn over the end? How long could I avoid acknowledging her? She might be crazier than a shithouse rat, but she was nobody's fool, and I hated the idea that she would look up and see me for the little Judas bitch I was. I hadn't for a second forgotten that I wouldn't be at Whiteman without her.

The boys had stopped talking about skis. In fact, nobody was saying anything at all, as Daisy's mumbling suddenly became audible. Hot sinking shame flooded through me as I promised the Virgin eighty-seven novenas on bleeding knees if she would just keep Daisy from really getting going. I shrank down in my chair and buried my head in my book, but I knew what was coming.

"Man drive the snowplow up the road. Knows good and goddamn well to lift the blade when he get to my gate. What he do? Block the son of a bitch with a pile of snow three feet high. Won't give a fella a chance. Can't stand it that a Negro has stayed right here in this one place thirty years."

There was a snicker behind me, Dale's. No one else might notice, but he was the kind who liked to lead a gang, even a small one. His previous victim, a boy with epilepsy, had left the school, and now he needed fresh meat. I knew how it worked. I'd seen a flock agree when a wounded chicken should be pecked to death. I had gone to great lengths to stay out of his way because I had been raised never to fight, never to act "niggerish." Now I had no idea how to protect myself from this bully.

Daisy now became transfixed by the yarn pills that had gathered on her blue sweater. She began picking them off, probing her breasts, pulling them up for better access. I put my face down into the textbook, inhaling the ink, the paper, as if I could escape into Side $2 = \sqrt{(hypotenuse)^2 - (side\ 1)^2}$. Around me the room was unnaturally quiet. I looked up. Daisy raved on. "I done the best I could. I done tol' you I caint do nothin with her."

I could feel Dale's cool cornflower blues on me. He snickered and whispered something to Ron, who snorted like a startled horse. Daisy, taking no notice, ran her tongue under her dentures and dislodged them to get at some irritant. The boys collapsed, laughing so hard they nearly fell backward out of their chairs. The other students laughed too, uneasily, uncertain what to do. Fury at Daisy, Dale, all of them coursed through me. There was no place to hide. The thing had to run its course.

Dale put his head next to Ron's and whispered. He said the word very quietly, but I heard it. "Nigger." I looked up. Ron's eyes widened and he glanced at me, astonished at first. Then he giggled so hard he fell out of his chair.

Dale grinned, metal braces gleaming, his eyes on me steady. I knew I should stand, find the Geronimo in me. But all I could do was stare back.

"That your mom?" he asked sweetly, twiddling a pencil back and forth.

"No, she is most definitely not my mother," I answered as archly as I could with tears welling in my eyes.

"Oh, yes, I definitely think that's your mom," he said.

~~~

My inability to speak up to Dale that day locked it in. He realized he could say anything he wanted, and I would return neither bark nor bite. Now he centered all his contempt on me, in study hall, in the van on the way to the mountain to ski, in English class, in geometry, in every way and at all times. I didn't expect anyone to stand up for me, and no one did. And in an eerie synchronicity, it seemed that Daisy hated me more and more each day as well, and expressed it in much the same way.

From Daisy: "Got your mouth all poked out. And I done give you everything. Well a nigger's a nigger and a mule's a mule, and you can't make nothing out of neither one of 'em."

And then on to school where Dale waited: "Here comes the nigger, with her double-double bubble buns."

Home: "Big as a bear, black as a crow. Talk more shit than the radio."

Dale, in study hall, whispered ostensibly to Ron as I walked by: "You're not a man till you fuck a nigger."

Home: "Niggers ain't shit. Caint do nothin for me."

One day, outside at the smoker's pit, Dale called me a name far more wounding than any he'd previously come up with. "Know what? You're a phony." How could I deny that? Because by then, when I looked inside myself, I found nothing. Every single day, for years, I had been trying to come up with a personality that would not draw the pack. But he had confirmed my deepest fear. Nothing I could ever do would make me worthwhile.

I began taking my lunch tray to the bathroom when there wasn't an adult around. Somehow it never occurred to me to tell one of them what was happening. It was as if by admitting what was going on I'd be like a wounded animal revealing itself, and I'd be devoured.

Somehow I hobbled through the rest of the term, but when graduation morning finally dawned, I didn't get up, even to feed Jiggs. I just lay there, so numb I couldn't even feel my body. When Daisy came back in from dumping the ashes, she was incensed.

"You still layin up in that bed?"

"I'm sick," I said burrowing as far down under the covers as I could.

"I have never seen such a lazy child in all my days. Get up out of that bed and put on some clothes."

"No," I said, and the house grew silent.

Daisy came to the back of the house and stood over me. "Are you sassing me?"

"No, I am not sassing you. I am telling you I am not going to go to their fucking graduation."

Even when she hit me I didn't move, and that scared her.

"I'm calling the sheriff," she said. But I knew Lowell was expecting her to clean the hall for the graduation reception. When she came back hours later, I was still lying in the same position as when she left. She stood over me again and let me know she'd had enough. I was going back to Perry Mansfield this summer all right, but it wouldn't be to learn all that mess I had before. All that dancing and carrying on had given me a big head, made me think I didn't have to obey. Made me think I was too good to work. That was for white girls anyhow, white girls with means, who had a mama and a papa to do for them.

Oh, I could holler and carry on all I wanted. It wasn't gonna do no good. And no, I couldn't ask permission to take even one class. Daisy had done got me a job in the kitchen with the cooks. I could peel potatoes and wash pots and pans, and when I was through with that, I could scrub up some toilets. It was time for me to come back down to earth. Otherwise, we could just haul my rusty-dusty on down the road to the reformatory. She'd be happy to call the sheriff to take me there right now. Now how would I like that?

I wouldn't, but neither could I stave off the despair I felt at being able neither to continue my dancing nor to see any other way out for myself.

It would have been so much easier if I had simply been able to tell myself that Daisy was lying when she said she wanted me to thrive. But the infernal conundrum of our relationship was that Daisy sincerely did want me to do well, and at the very same time she didn't. Very dimly, I connected this contradiction to the world of complete confusion into which she had been born—every man created equal, except if that man was black, or not a man but a woman. None of it had made any sense then, and neither of us had the tools to sort it now.

~~~

Ironically, it turned out that I would be working not at Perry Mansfield but at Whiteman. Portia and Kingo had so many students this year that they needed extra space, so I would get to serve my old friends from Perry Mansfield in my recent torture chamber. I dreaded running into anyone I knew from either place. Would they pity me? I couldn't stand that. But I was losing contact with the things we had in common. What would we talk about—the best way to get pubic hair out of the caulking behind a toilet?

Every morning I woke in a fury. Dragging myself into Whiteman, I made a point of keeping my eyes on the floor so I wouldn't have to see the empty study hall or the vacant class-rooms. I stood outside the kitchen door to collect myself before I entered, listening to Peeler sing the spirituals that always left me at a loss. Catholics didn't do spirituals, and we weren't much on joy. But Peeler marched into the kitchen every morn-ing with her snow-white uniform ironed, her sleeves cuffed so they would stay perfect all day long. Her apron could have stood up and sung "The Battle Hymn of the Republic" on its own. A tall, thin woman, with dark skin and regal bearing, she looked quite formal with a black chignon at the nape of her neck. At Perry Mansfield, she had been a "salad girl," but here at Whiteman she ran the kitchen, and she made a point of run-ning it immaculately.

When I opened the door and walked in, I could never hide my long face. Peeler looked baffled at first, as if she had no more idea than I did exactly what I was doing there. And then, as she began to contend with me hiding and half doing the work and generally being a pain in the behind, she began to look annoyed. She didn't need a sullen teenager to babysit. How could I ex-plain that I was dying inside—that lately my only comfort was the thought of truly dying, and that only the thought of aban-doning Jiggs troubled me when I considered it?

Then Peter showed up. A great dancer from the previous year who had come to Perry Mansfield on a work-study schol-arship, he was not only handsome but graceful and funny, with wavy hair and huge blue eyes. I had admired him from afar, but now, working alongside him in the kitchen, I got to know him. I lived for the moment when he walked into the dank back room where the sinks were. He brought the news—who had come back this summer, who had graduated to *pointe*, what the productions were going to be. He was pretty much the only person under what should have been retirement age that I interacted with each day, and in short order, he became my best reason for living.

Peter only worked the lunch shift. We stood at the big sinks together, him washing, me rinsing and drying the big pots and pans. He made everything fun, doing outrageous leaps and jetés with soup ladles and colanders as props. He told me he was going to Stephens in the fall, or maybe New York. How talented he remembered I had been, how hard I had worked at my turnout. "Why don't you take Harriette's class with me next week? She's doing a master class, and it's just going to be fabulous." I had never heard anyone use that word before, and immediately adopted it for my own, telling Jiggsie every evening when I brushed him that his coat was "simply fabulous." I didn't bother to tell Peter that Daisy wouldn't let me take a class. I knew in the floor of my soul that even though every noon Peter and I stood in the same place, I had deviated ten degrees south. We were not heading in the same direction at all. In another year, we wouldn't even recognize each other. I would never be a dancer or any kind of actor, but he would.

Once again, the end of camp came, and people were packing, all in a rush. One of the costume designers loading up his car handed me a bottle and said, "Give this rum to your aunt. She can use it in her fruitcakes." I was helping with the cleanup, and I had told Daisy I'd find my own ride home, and so it seemed the most natural thing in the world to ask Peter if he would walk me home in the moonlight. We'd sip a little rum, get to know each other. I knew he adored me. And I loved him.

We walked down the road. Aspen leaves shivered in the warm, breathlike air, delicious with promise. I took a sip of the rum. It tasted like kerosene, and just like fuel oil, it turned on a peculiar heat that cheered me considerably. My troubles with Daisy, my future, seemed like no big deal. The only thing that existed was that August moon dressed in a gown of gauzy silver, swaying in that ballroom sky. And when I looked up at Peter, he had become my cowboy hero, tall as an Apache brave.

"I love you, Peter, and I will for the rest of my whole life." He was walking too far from me to touch, but he heard me, and he smiled.

"You too, Rita. You really are truly fabulous."

I couldn't help it. I galloped across the road and flung myself on him, mashed my drunken lips against his, and waited for him to take me in his arms. Instead they hung at his sides like wilted celery, and suddenly the moonlight shone way too bright. We were both embarrassed, and he turned away. He tried to hide it as a shrug, but I knew he was wiping his mouth with the back of his hand.

Why had I ever thought anyone would care for me? Thank God for the booze. I took another swig and slunk off.

"Don't be mad," Peter called. "I really do love you to pieces. Just not *like that*." "What's that mean, 'not like that'?" I asked. But I knew. Every single rejection I had ever had piled down on me like an avalanche. Nobody would ever love me. Nobody ever had.

"I love you like a sister," he said. "Or a brother." Then it dawned on me. Jesus, I had the worst luck. We went along quietly for a while, the gravel grinding under our feet. It was going to be a long walk home.

Finally I thought of something to say. "You know what? When we first heard about sodomites, my friends and me thought they were some kind of mite or earwig, or like a tick. Isn't that funny?"

Peter said nothing, but I was undaunted.

"Oh, I don't think that now. That was just when I was a baby and they talked in church about sodomites in Gomorrah. I thought they were just some kind of bug from biblical times."

We were walking down a stretch of road that was ringed with tall pines, and I couldn't see Peter's face very well, but I could feel that the evening had taken a downturn. Something told me to shut up, but I couldn't seem to heed it.

"It really was my friend that thought that. It wasn't me." Why didn't he say something? His strides had lengthened, his dancer's legs so long I could hardly keep up with him.

"Peter? You want something to drink?" He looked back at me, his thick hair highlighted by the moon. He looked like a character on stage, lit to emphasize shadow. At least he reached out and took the bottle.

"Listen, my friend didn't know anything. She thought it was like when sheepherders hobble together the back legs of their ewes and put their front hooves in a bucket so they can't get away, and then they stick their thingy in them." That struck me as funny. I laughed so hard I fell down. "Best lock up the henhouse," I gasped. "The rabbit hutch too." I was giggling so hard I nearly peed my pants. But Peter kept going.

"Wait," I croaked. He didn't even turn around.

I finally staggered to my feet and lumbered after him. "Look, don't get all mad. It's just a joke, okay?" I reached out to grab his arm and he nearly decked me. The moonlight revealed the face of a really angry man.

"Look Rita, I thought we were friends. I didn't realize you were a total country bumpkin. Now leave me alone." And this time he walked off for real.

"Peter, come on." I couldn't believe that once more, my reason for living was disappearing down the road. "I didn't mean you'd fuck a sheep. I know you better than that."

"Rita, just go home."

I sat down in the middle of the road. The gravel was cold and sharp. "I'm going to stay right here until you come back," I called after him, but he kept going. I started to cry. "It'll be your fault if I get run over," I yelled, desperate now, but he crossed down past the turnoff and kept going.

Now what? I'd promised I'd stay right here. It occurred to me that I was a little drunk. Well, I might as well get really plowed and make a proper mess of it before Daisy shipped me off to the pen.

Next thing I knew I was looking up at the grille of a vehicle, its headlights peering down at me. The engine smelled like it needed an oil change.

"Rita?" A woman's voice. Oh, shit. It was the new neighbors who had moved into Daisy's upper cabin.

I sat up, bemused by a lovely stream of silver snot trailing from my nose to the dirt. I tried to get up, but slipped and fell on my face. Then somebody's arms were hoisting me.

"Come on. You've got to get out of the road. Let's get you home."

I was way too drunk to care where they took me, so it was over the hill and through the dale to Daisy's house we went, and soon we were outside the gate. Daisy had left the yard light on. I got out and threw up in the ditch.

Daisy opened the door. "Reeter Ann?"

Mercifully the neighbors just drove off.

"Hi, Daisy, it's me." Jiggs was whining. I tried to make my way to the door, but I tripped and fell flat out on the bed of peonies.

"Sorry." I tried to get up like it was nothing.

Daisy's arms were crossed. As I came near I saw her nostrils flare.

"You've been drinking?"

I shook my head. "Of course not. Where would I get anything like that?" I had to hold the doorjamb to steady myself.

"All right, that's it. I'm going to get the sheriff."

She got her keys and brushed past me down the stairs. I was still standing in the same place watching her one red taillight go down the hill, thinking I should have told her to fix the other one before the sheriff gave her a ticket.

I figured I had just enough time to blow my head off and be done with it before she got back. I found Gampa's .22 under the buffet, but I couldn't find any bullets. Then I found the straight razor he used to shave with. I'd cut my throat. I searched my neck in the mirror, but I had no idea where the goddamn jugular was. And then I heard poor Jiggs jumping up and down, barking and rattling his chain. If I didn't do it that instant, I'd lose my nerve.

Then I remembered. Somebody, somewhere, said the wrists were faster. Was that true? At least I wouldn't have to look myself in the eye. I drew the razor across the left first, and then the right. My arms opened and mercifully the pain went there, because I couldn't stand the other pain anymore. But I hadn't imagined it was going to hurt so much and bleed so hard so fast.

Jiggs was going crazy. I had to get to him. But there was lead in my feet still, and I tripped over the piano bench and hit my head on the television. Finally I got to the front of the house. I knew I had to get out before Daisy got back with the cops. I fell down the stairs and then remembered I had to find Jiggsie's key. Had Daisy hidden it in a new place again? I dragged myself back into the house. The key was still under the telephone book, but I was bleeding too hard to hold it. Maybe I cut the ligaments too deep. I put the key in my mouth and fell back outside. Jiggs was absolutely wild. He bayed like a hound and wouldn't hold still, and the key was rusty and there was too much blood, running black in the moonlight down into the coal dust next to his side of the porch.

I finally got the padlock to pop, but Jiggs kept getting under my feet and the chain tripped me up. I fell down again like a load of timber and landed on Jiggsie, who yelped. I got to my hands and knees and saw that I had thrown myself against the chopping block, where the ax was impaled like it was waiting for the right neck.

We had to get out of there. I wanted to go out across the field, but I couldn't face the barbed wire, so we headed north. I was freezing, and I couldn't figure out why. I got as far as the Swigerts' barn, where I thought maybe I'd take a little rest. Jiggs wouldn't stop licking my face and whining. I wanted to hug him, but my wrists hurt too much to move.

This time the posse found us right off. It wasn't a whole mess of guys on horses, just two average-sized cops in a Jeep. "Don't take me to jail," I begged. "Please don't take me to the reformatory. Please don't put handcuffs on me. Where's Jiggsie? Where's Jiggsie?" I was drunk and disgusting and I couldn't find my dog.

"What made you do this, sweetheart?" one said. Now what was I going to say to that? I wanted to die. What did he think made me do it?

They put me in the Jeep and took me to the hospital, where the light was hard, blue as old white people's varicose veins. The emergency room had cloudy glass walls that smelled of Windex. I wondered if I should tell them it was cheaper to clean windows with vinegar and newspaper. I sat on a cot behind a curtain, the maw of my open wrists odd, out of context. But what would be the proper context?

The bleeding had slowed, and it looked like I had made a mess of this too. Apparently I was going to live. A nurse with a cartoon hat came in and arranged steel stuff on a tray. Then she gave me a quick pat and scooted out.

Outside the curtain, I recognized the voice of the doctor who had known me all my life, from my first strep throat to the day Daisy had hauled me in so he could tell me about menstruation. I stank of puke and rum, and the clotting blood smelled like a rusty iron file.

Dr. Richards stepped through the curtain with squeaky shined shoes. He switched on a bright hot lamp on rollers, sat on a stool, and wheeled himself over to me. I was so ashamed. I couldn't even die right.

"So, Rita, what have you done to yourself?" How come he put it like that? It just happened. I could feel him willing me to look up. At least I could control my eyes. I studied my torn jeans, only a month old. Daisy would be mad. I couldn't figure out the categorical syllogism that would explain how I had wound up here with my arms gaping open like a gutted rabbit. Daisy and I both thought an education was going to fix me. What would Aristotle say about this goddamn mess?

Dr. Richards reached out and took my arms, brought them up, turned them over to examine my wrists. "You used to play piano?"

I nodded.

"Don't know if we can save these tendons or not, but either way you cut through a lot of nerves. I doubt you'll have any feeling in your wrists again." Oddly, I was more upset at the idea of losing the feeling in my wrists than at the idea of losing my life.

He fiddled with the light and brought a magnifying glass down over the wound. His aftershave made my nose tingle. He took out a huge needle and said, "This is going to hurt, so brace yourself." I laughed. He didn't. When he pushed it into the ragged edges of the wound and pressed the plunger, I screamed. It was like hot ice. He patted me. "There. It's done. Now we can stitch these up."

"So, I'm for sure not going to die, huh?"

He looked at me with surprising openness. "Sorry."

I kicked the bed. It startled him, but he stayed put.

"You know, before you really get serious about this business we should get you some help." But I knew there was no help for me, so I said nothing.

"Dunno if you'll be able to write with that right hand. Might have to become a lefty. The nerves may reconstruct, but that could take years."

And then he started sewing on me like Gama would stitch up a pillow tick. How odd. Only twenty hours ago, I'd been picking raspberries. Now I wasn't sure I could pick my own nose.

They gave me some pills and put me in a pretty room all by myself, lots of blue and white with a potty and a sink and a toothbrush I couldn't hold. I could feel the medicine pushing me down toward sleep. My final prayer was that Jiggsie be okay.

The hospital had lots of night spirits that clicked, hummed, squeaked, and sighed—maybe that was what healed people. Somebody down the hall was snoring, louder and louder, until he woke himself up with a snort. Then it was quiet for a minute until he started up all over again. My sheets were hard, and smelled as if they'd been boiled in bleach. The rubber mattress made me feel like I was going to slide out the window and down the hill.

And then a cart rolled down the hall and woke me up. In spite of everything, the dawn arrived cheerful and buttery. With me or without me, the world had a life.

I could hear kids playing hopscotch outside. I wished I could see Lynne or Marie or Janet, my old playground friends from Steamboat, or Tom and Jill from Whiteman. But I knew Daisy would die if anybody found out about me here. I lay there, waiting for her to come marching in with the sheriff. My wrists had started throbbing like there was an engine inside them. I wondered how they'd get the handcuffs over the bandages. My middle fingers didn't exactly work, but I could pick up things pretty well with my thumb and pinkie. I couldn't seem to slake my thirst, but every time I drank water it made me drunk all over again. Even on the fourth pitcher I had to hang my foot out of the bed and touch the little footstool to get the room to stop swirling.

At lunch, the fish sticks and canned corn made me cry. They reminded me of grade school when Lorene Workman made pigs in a blanket and Mary was still home and Gama and Gampa alive. The only thing I wanted was the fruit cocktail, with its single watery half of a maraschino cherry. My wrists throbbed and twitched like they were trying to knit themselves back together. The doctor didn't come, just the nurse, to put a thermometer in my mouth. Down the hall a radio played—Marty Robbins singing "out in the west Texas town of El Paso" on KRAI, the station Daisy always listened to. That day nobody came to see me. I guessed I was glad.

Next morning Dr. Richards swooped in without knocking, catching me not exactly decent. "So, how are we doing?" he asked. I didn't exactly know what to say to that. Did he mean how was *I* doing? He put a stethoscope to my chest and had me stick out my tongue.

"So when can I go?"

He made a big to-do about taking out a little flashlight, putting a cuff on it, and sticking it in my ear. He didn't answer my question for a full count of ten as he fiddled with my other ear and said, "Put your head up." He stuck the flashlight up each nostril as if I had been hospitalized for a head cold.

"Well, Rita, we'll have to see." There it was again—that *we*. "Let's take a gander at those bracelets."

He unsnipped the bandages, revealing arms that looked as if they belonged to Frankenstein's bride. Twelve stitches on the left, ten on the right, where I had tried to open the seam and let the darkness drain out. The little knots were black with dried blood. "Looks pretty good, kid," said Dr. Richards, and then he was gone, leaving an eddy of doctor-shaped questions in the room.

Nobody else came that day or the next except cheery nurses with meals, and I started to think that Daisy was up to something heavy.

On the fourth afternoon, I was dozing in the warm sun, wondering whether I could try flexing my wrists. I thought I heard something. Maybe it was the dinner cart coming with the pork and beans and weenie that was on the menu. When I felt someone in the room, I opened my eyes.

Daisy stood at the foot of the bed, studying me, just like she used to peer in at me through my window. She had on a clean bandanna and a crisp blue denim shirt. The Jeep keys jingling against the metal bed rail must have been what woke me. For about two shakes of a lamb's tail I was glad to see her, and then the dread surged in.

"So. Are you satisfied? Embarrassing me in front of the whole damn town?"

I tried to sit up, putting my weight on my hand and stretching the stitches painfully. She came around the edge of the bed and looked down at me, her jaw muscles flexing.

"After everything I have done for you. Treated you just like you was one of my own." There was that inescapable Daisy logic again. My cut wrists were a slap at her. Still, I was ashamed not to have done it right if I was going to try it at all.

"I'm sorry."

"Bought you Kotex and pencils. Took you to piano lessons and the Bluebirds. And now you lay up in here and stare at me like a goddamn bay steer?" I never had understood what she meant by that, and I wasn't sure where else to look. I couldn't think of a single thing to say, so I didn't say anything.

"Negroes don't cut on themselves, Rita Ann Williams." Daisy threw her hands up and stomped over to the window. "Precious Jesus, I never thought I'd say this. But I am glad Mama and Mae is in the ground dead so they don't ever have to know any of their kin ever did anything so ig'nant." She whirled around. "Answer me, heifer."

It didn't seem like a good idea to say, "What's the question?"

She turned back to the window, pigeon-toed and slumped. How could she be all stove in and irate at the same time?

"Dr. Richards thinks you gone mental and say he fixin to ship you off to Pueblo." I felt a stab of betrayal.

Daisy opened her handbag and threw a pair of jeans and a shirt at me. "Get up and put these on. I'm takin you out of here."

I noticed she had chosen a long-sleeved shirt for me. It hurt to work the buttons. She watched me dress, the waning light illuminating one side of her face. The other, like her personality, was in a shadow I could never see into.

After I had struggled into the jeans, she tossed me a bandanna, red like hers. "Cover that mop."

When we went out into the corridor, there was nobody at the nurses' station. "All right," she said, and I wondered if she was breaking me out without Dr. Richards's permission. For once I was glad to go with her. I didn't feel like sticking around so he could cart me off to the loony bin.

To my surprise, Daisy didn't turn the Jeep toward home but headed toward the main street instead. "Now. We are going to drive through town, nice and slow, so everybody can see it's a damn lie that you tried to take your own life. You smile at every-body and look pretty." I wanted to sink through the springs of my seat. "Sit up sharp, Rita Ann. Don't make me have to tell you twice." I did as I was told.

We drove all the way from the east end of town to the west, but we didn't see anyone we recognized. It occurred to me that maybe no one was all that concerned with us, but I knew if I said that she'd get mad all over again. At the post office, I was certain she was going to make me get out and go in with her to check the mail, but she didn't. The she stopped by the *Pilot* to get the paper. I felt very peculiar. I began to wonder if the other night had been a dream, or if I was dreaming now.

As we waited in traffic to make our turn home, the door to the Pioneer opened. I heard a short burst of country music and then a man wearing a cowboy hat set way back on his head held the door for a girl in very high red heels. I recognized her. She'd been two years ahead of me at Steamboat, and she'd filled out, as they say. She'd gotten her mousy hair ratted up in a beehive too. She slithered toward the man, then grabbed his arm and began to laugh, first with her head thrown back, then doubled over forward. I watched until we turned the corner, thinking, here's a girl acting like a filly in estrus for the whole town—the whole county—to see, and she's not dead.

~~~

At home, Jiggs was chained and padlocked again, and Ernest was visiting. He didn't look at me—he never had, but this time it was deliberate. Daisy made fried chicken and greens and cornbread, and gooseberry pie for dessert.

"I think these onions'll take a blue ribbon at the fair," Ernest said, holding up a huge purple slice to the window and turning it from side to side. "White ones look pretty good too." I started to wonder if I he couldn't actually see me.

Daisy offered him the platter. "Ernest, you want the back?" My favorite piece.

He took it, but he didn't eat it, just tore it apart and lit a cigarette.

My wrists itched like hell, my only evidence that that soggy night had ever taken place. But the chicken was perfect—the crust crisp and garlicky, the flesh moist. When we were done, only the breasts were left.

"It's a funny thing," said Daisy. "Don't nobody want the breast, and that's all the white folks ever eat."

"Yeah, I don't get it. It's so dry," I said. Ernest glanced at me, then looked away.

Daisy settled in her red chair. Time for the *Pilot*. She read in silence, absorbing every page. Ernest set aside his almost untouched plate and fell into the Wayside Gardens catalog. Both of them dropped like stones to the bottom of a lake. I remembered what it was to read like this too, leaving my body and taking on the cloak of the print.

I cleared the plates awkwardly—the middle fingers on my left hand were very stiff. Bending down low to the coal bucket to eat Ernest's leftover chicken, I chewed up all the bones that weren't too big. I tore off a little piece of breast to make a treat for Jiggs. The minute I shook his dinner bucket, I heard his chain rattle. He'd love Ernest's buttered cornbread and milk gravy. I poured the rest of the chicken grease over the bones.

"Don't give the dog that gizzard and liver, and don't touch his chain," Daisy said, turning the page but not looking up. I took them out of his bucket, but decided to leave the feet and head and neck on the counter. I couldn't face peeling the yellow skin from their little limp toes and pale curved toenails. Then I grabbed the coal bucket and pushed the handle high up my arms past the bandages, so I could "make myself useful."

11 ~ BVM

Me, dancing

~~~~~ **Two weeks later,** early in the morning, I was outside in the berry patch cutting up cheesecloth to cover the strawberries when I just couldn't stand the way my stitches pulled anymore. Daisy had not once brought up the incident since I'd been home, and I was still leery of Dr. Richards sending me off to the nuthouse, so I decided I'd snip them off myself. Why not? Daisy had once told me how Mr. Anderson had gotten a bad cut and stitched himself up.

When I was finished, the wounds ached. They looked pink and angry, as if they were going to keloid. I flexed my hands back and forth, wondering whether I was ever going to regain all the feeling in my fingers. I decided to go into the house and get some tape to cover my wrists.

As I came up the stairs, I heard Daisy on the phone. "We'll be there this Saturday," she was saying. Adrenaline pumped through me. Was this about me? I had been a model teenager since I'd been back home. Made my bed before I even showed my face every single morning, and I hadn't once tried to find the key to Jiggs's padlock. But Daisy had given no indication that she took note. In fact, she hardly looked at me at all.

She laid the sticky receiver back on its hook and brushed some specks of onions, dill, and carrot off with her sleeve. The phone must have rung when she was mixing up a crock of America's relish. I went to the piano and tried to play "Chapel Chimes," but my left hand, never my best, stumbled at the octave. I made it to the end, though, and glanced toward her. She was sitting at the table with a pile of bright red peppers and cabbage. She'd brought out the whetstone, and carefully dropped oil on its rough side before she honed the paring knife, her favorite, which she'd already worn pencil-thin.

"The sisters out to Boulder gon' take you on. I told them I believe you want to be a sister too."

Tears burst from my eyes. "Oh Jesus, Daisy, I don't want to be a nun. Please. Please." She wouldn't even look up at me. Instead, she carefully drew the narrow blade across the stone. If I couldn't make it as a student at a Catholic school, how was I going to convince anyone I had a vocation?

"Daisy, don't. I'm begging you." She rose and padded slowly back to the cabinet, where she stood on tiptoe, moving bottles from side to side on the top shelf until she located the cider vinegar. "They done studied your Whiteman grades and fixin you up with a scholarship." She measured out a fistful of pickling spices and tossed them into the stone crock she'd propped on a chair. "I told them you was plenty smart."

"But Daisy, that doesn't mean I want to be a fucking sister."

That got her attention. "You'll not raise your voice to me, nor disrespect me in my own home. I done done all I'm fixin to do. Now that's the end of it." She didn't even sound angry, and that chilled me more than anything.

Saturday afternoon found me once again waiting for the bus, like a load of freight that wouldn't be claimed. With the tourists gone, the town seemed drowsy. Toward Storm Mountain, the fall leaves made it seem as if a tapestry of gold and evergreen had been hung above us. When we saw the coach coming, Daisy opened her pocketbook. "Here's ten dollars. I'll send more when I get it."

I didn't recognize this bus driver, but Daisy did. "Make sure my gal catches the right bus to Boulder when you all make it to Denver." She had never called me "my gal" before. She reached out and gave me a stiff hug. "Honey, this is your last chance to make something of yourself," she said. She looked so tired in her run-down shoes and faded plaid shirt. Before I could think of what to say, she turned and trudged back to the Jeep.

In Kremmling, the bus made a fifteen-minute stop at a coffee shop. I waited until the driver and the rest of the passengers were off, then I unbraided my hair, loosened the collar of my shirt, and got off too. Through the restaurant window I watched the intersection. What if, instead of getting back on the bus, I went to the corner and stuck out my thumb for the right cowboy in a wide-body pickup? When I turned around, however, I noticed that the bus driver hunched over his apple pie at the counter was studying me under the brim of his cap. I waited till he went to the john, then bought a pack of Pall Malls and some breath mints, but that was as much adventure as I had the stomach for.

Back on the bus, I slept, soothed by the familiar engine, tucked into three safe hours of predictability. But I dreamed about Jiggs, and when I awoke, I was pierced with sadness thinking about him chained under the porch next to the coal. How he hated having to do his business five feet from where he ate. And it would not occur to Daisy to set out planks so he wouldn't have to lie on the frozen ground or brush him when his winter coat came. She wouldn't think to bring him warm water when his had frozen, believing he could just eat snow.

I studied the mountains I was leaving as the bus groaned over the passes. Now that I was older, the trip seemed shorter. I made the switch in Denver, and half an hour later, Boulder with its spectacular Flatiron cliffs came into view. It excited me mildly to be in a college town. Colorado University was known as Party Town U. Not that I'd get a chance to find out for myself.

Before the driver even engaged the air brake, I had picked out the nuns' car. Sisters always favored pale blue station wagons. There were two of them under the awning, looking earnest and expectant. They were tall, and with their peaked headpieces and flowing habits, they looked like little mobile churches. The minute I stepped down from the bus, both of them looked in my direction and beamed.

"You're Rita," the tallest exclaimed. "I am Sister Mary Regina, and this is Sister Mary Angelice. We are absolutely delighted to have you."

After the send-off Daisy had given me, I wasn't prepared for this brass band. What had she told them? I pulled my sleeves down carefully, hiding my blood bracelets.

Mount Saint Gertrude Academy for Girls didn't sparkle like Mount Saint Scholastica. Only a short distance up the hill from the station, the four-story red brick building penned in by a severe black iron fence looked like a jail. I wouldn't be sashaying out of here to have a chat with any motel owner. But as we made our way to the entrance, I heard someone playing the very Rachmaninoff prelude I'd been wanting to learn, with skill and feeling.

"That's the conservatory," said Sister Mary Angelice, indicating a small white house adjacent to the main structure. I wiggled the fingers of my left hand tentatively. Could I actually study piano again? What if the teacher saw my scars?

Sister Mary Regina held the door open for me, beaming down at me as if from on high. "Welcome," she said.

A statue of the risen Jesus dominated the entryway. Oiled burl, dark and coiled, it made me want to turn right around and bolt. But I willed my body to follow Sister Mary Angelice's swishing skirts, old-lady shoes, and huge clicking rosary up three flights of stairs. We came out on a wide gray tiled landing that had been polished like glass. Conversation and peals of laughter came from unseen rooms beyond.

From my room emanated what sounded like a party in full swing. "Girls," Sister Mary Angelice called out as we entered, "welcome our new student, Rita." We stepped into a large white room with wide windows through which I could see the pretty city of Boulder below. Ivory curtains hung from metal pipes at ceiling level, dividing the room into five private alcoves. Inside each was a cot and a small dresser. "Help Rita get settled in by first assembly, all right, girls?" said Sister Mary Angelice. And after regarding me with a generous smile, she was out the door with a swoosh.

"So you get the sixth bed, eh? I'm Janet from Fraser," said

a stout girl with hair dyed the shade of a hard-boiled egg yolk and a stiff body odor. I found it curious that no one had addressed her about either. "Put your stuff here," she said, pointing to the cot. Oh good, I thought to myself. Bossy girls could be tiresome, but from experience I knew they were the ones who could tell you where the shoulder dropped off the road.

"I'm Marsha," said a short, round girl talking around a wad of purple bubble gum. She blew a balloon-sized bubble and smacked it with the flat of her hand. When she realized she had hit herself in the mouth harder than she meant to, she burst out laughing.

"This is Elise," Janet said, pointing to a languid blond lounging on the corner cot, twiddling her hair with her long manicured nails as if she were posing for a lingerie ad. Well, Daisy, I thought, you finally got me corralled with a fine herd of happy heifers who never heard of the open range.

I had trouble sleeping that night. The hall lights stayed on and the windows were locked, making it feel as if the very air was cooking. Outside the sky glowed orange, as if a forest fire was smoldering close by. Eventually I realized that Boulder's mass of electricity must be keeping the dark at bay. I put my pillow over my head, but that didn't help. Every time I turned over, the springs of my cot squeaked. Around me the other girls snored softly. If I reached out my hand, I could touch Marsha's cot. I missed the weight of Gama's quilt, but when I piled on my winter coat, I got too hot.

I slipped down the hall to the drinking fountain. I couldn't get the water to run cold, and it smelled as if the pipes were lined with Clorox. I went back to my cot and listened for the building's voice, for its soul. At home, our wooden house clicked all night, through every increment of the temperature dropping. But this brick place had settled long ago, as if it were a catacomb. With all the curtains closed, there was a mysterious sense of aloneness, even though I could hear every breath my sleeping sisters took. Restless, and missing Jiggsie keenly, I kept tossing long into the night.

In chapel next morning, there she was—the exact black girl Daisy had always wanted me to be. Head bowed devoutly beneath the little lace doily bobby-pinned onto perfect hair, with every trace of kink reconstructed into unlikely coils, she glowed in the morning light, her skin the color of a root beer float. The Peter Pan collar on her buttoned-up blouse covered a gold chain that revealed, at the throat, a beautiful gold crucifix. When she caught me gawking, she smiled graciously, revealing a mouth full of expensive wire. Then she returned to her devotion with a sincerity that couldn't be faked, closing her eyes in prayer.

My heart sank in profound confusion. I bet this girl had never heard the word *nigger*. With her little white missal and her perfect little do, she seemed at first glance as white as any other Mount Saint Gertrude girl, maybe more so. Later, at breakfast, I found out that her name was Maple. She had saved me a place next to her. "It's so good to see you," she said after grace. I wondered exactly what she thought she was seeing. "Did you come here because you have a vocation?" she asked sweetly. Against my better judgment, I nodded, hoping that if I tried to inhabit the role, I could make it fit.

Next to her plate, she had a letter ready to mail to her parents. "I write them every day," Maple said. Her handwriting sloped uniformly, and she had made little hearts over her *i*'s.

The day passed in a blur. Introduction, registration, picking up uniforms, buying books, and finally out to the conservatory, where I met Sister Mary Paul.

At dinner I tried to find someone else to sit with, but Maple came and tugged at my sleeve again. "Let's sit here," she said, pulling me to the table headed by Sister Mary Regina. I desperately wished I could pour my soup into a can and drink it in the restroom the way I had at Whiteman. But Maple seemed at ease. "Bless us, Oh Lord, and these Thy Gifts." She mouthed grace along with the mother superior, then afterward, keeping one dark hand on the napkin in her lap, she dipped her spoon into her broth and carefully scraped it against the back of the bowl before she took a dainty sip. With my dead hands, I didn't even bother trying to wrestle the soup to my mouth.

Next morning a fully dressed Maple passed me as I was on my way to the lavatory. She looked at me, concerned. "Are you all right? You didn't finish your dinner." There was no getting around it. Maple was cheerful and sweet, generous and kind, and the very sight of her made my teeth itch.

I felt utterly hemmed in at Mount Saint Gertrude, although no one had been remotely unfriendly to me. Like a caged bird slamming against a windowpane, I longed for open sky, the gentle murmur of trees, the call of owls at dusk. And boys— their sideways glances, their big paws. Through the building's tiny windows, light did not shift in the gradations I was used to. The fluorescent tubes overhead flickered constantly, staining the gray walls a cold, jumpy blue.

Also, I was never alone. At 5:30 in the morning, we were cheerfully herded to chapel. From 6:00 to 6:30, hygiene in the community bathrooms. Then prayers before breakfast, and breakfast, after which I washed dishes as part of my payback for tuition. When classes started at 8:00 that first morning, I nearly ran another student down trying to get a desk by the window.

My nose dried out in the radiator-heated air, but there was little I wanted to smell anyway. The girls camouflaged their own smells behind perfumes with names like Lily of the Valley, and the nuns smelled only of dust. The essence of anything more earthy and authentic had been eradicated here generations ago. Outdoors there were no skunks, no horses or hay, not even the familiar smell of moisture before the rain. I missed particularly the sweet breath of geese and the fragrance of hens setting on warm eggs. Here, the chickens arrived wrapped in plastic, devoid of the feathers, feet, and heads that I had hated so much back home. Now their absence unnerved me more. I couldn't believe how much I missed the lights, sounds, smells, and sweet mess of Strawberry Park.

Choir practice with its sad Gregorian chants echoed the grief I was still trying to outrun. It irritated me no one else seemed dragged down beneath the weight of this music. Every ounce of waking energy I had went to pushing past my sorrow, yet around every corner lay statues of Mary in anguish, the stricken disciples, and Jesus, in agony for my sins. It was enough to make me want to climb the bell tower and leap.

But one night, as I lay on the cot in my alcove, I discerned the unmistakable aroma of tobacco coming from the alcove next to mine. Marsha had been smoking. Next morning, when we were brushing our teeth in the bathroom, I looked in the mirror and said, "So, where do you go to light up?" Her big blue eyes popped open wide. She checked for feet under the three stalls behind us. When she was certain we were alone, she broke into anxious giggles. "How did you know?"

I didn't bother explaining. "I smoke too."

She brightened considerably. "Meet me behind the conservatory after speech."

I soon discovered that not only was Marsha afflicted with the same restlessness that bedeviled me, but with two years at

Mount Saint Gertrude behind her, she had road tested no end of schemes to get off campus to drink, smoke, and meet Colorado University boys. Her primary escape route was a hole in the back fence. We would have been easily visible from the bell tower, but I couldn't resist. Soon, rather than going straight to the conservatory to practice after classes, I was traipsing off down the alley behind Marsha. We even made friends with a lonely old German shepherd, who barked at us fiercely until I brought him part of my Friday fish-stick lunch.

"Saturday we can go down to the Tool and drink beer," Marsha announced at the end of the week.

"The Tool?"

"Tulagi's. There's always boys willing to pay for a pitcher. All we have to do is be back by five. And bring a toothbrush along." She had it all figured out. The first time I went with her to Tulagi's, a couple of football players, solid as steers, did indeed buy us a pitcher. We got back on time, although Marsha nearly tripped over the kneeler at vespers, sending herself into a snorty giggling fit. I found myself wondering what kind of warp in the universe I had just slipped through, and I knew it couldn't last.

But it did. Never mind the weight from a thick layer of guilt and surreptitiousness that was piling on top of the miasma that had nearly killed me in the summer. Or the overwhelming course load I signed up for—Latin and advanced algebra, American and Church history, piano, English, debate, and social studies. I couldn't understand why I couldn't quite get on top of my studies when I had done so well at Whiteman, not acknowledging that something was keeping me from engaging on anything like the same deep level. I got two humbling Cs on algebra midterms. Translating Julius Caesar from Latin into English seemed a pointless exercise because Roman conquests didn't interest me the way the capricious, philosophical Greeks had. I wrestled with my concentration, but I kept losing the thread and tuning in somewhere down the road, only to realize I had missed crucial information. Dimly I sensed that I was struggling to keep the lid on a cauldron of seething proclivities that was bound to boil over sooner or later.

One afternoon Sister Mary Angelice was going on about the fall of Rome when I thought I saw a snowbird land on a pine bough outside. I waited for it to reveal itself, but it remained hidden behind the trunk of the tree. "Daydreaming again, Rita? Let us know when you are available for class."

Shit a mile! I'd done it again. "I'm sorry, Sister."

Thirty sets of eyes had fastened on me.

"Well I'm afraid, young lady, that's not quite good enough. Would you care to answer the question?"

"Yes, Sister, 476 A.D." It was just a shot, but I lucked out. How, I didn't know. It happened in math too—sometimes I would get the right answer, with no idea how I'd done it. Sister was not fooled. She knew I had gone fishing. Still, she nodded and turned back to the blackboard. I had been just present enough to scrape by this time, but my treads were wearing thin. Especially now that I sat in the back of the class with Marsha and Dolores Smith, passing notes half the time. I felt like a slab of snow poised to avalanche, so unstable that a sneeze could dislodge it.

Now when I ran into Sister Mary Regina, her cheery greetings had turned to flinty scrutiny. In chapel I prayed, *Oh God in whom I do not believe, please don't let me get kicked out of this place I hate.*

One wintry afternoon, during the last class of an interminable gray day, Sister Mary Angelice asked the algebra class, "If a train is going thirty-eight miles an hour and has left the station eighty-nine minutes before a car that is traveling forty-six miles per hour, how long will it take for the car to catch the train?"

In the back of the class, the part of me that wouldn't be housebroke reared up and whispered to Marsha, who sat next to me, "Who gives a shit?" She slumped down next to me and began to giggle, then tried to muffle the sound by burying her head in the spine of her math book. The hysteria jumped the aisle to Dolores Smith, who slid down into her chair snorting like a pig. I realized with delight that I had found the wrong crowd. Dale's steel smile flashed before me and I mused, *So this is how it's done.*

I risked a peek around the girl in front of me and caught Sister Mary Angelice's steady blue eyes looking dead at me. Immediately I sat up straight, exerting every piece of iron I had not to glance in Marsha's or Dolores's direction because I knew that would set me off again. Sister didn't address the noise or interrupt the flow of her lecture, and I dared hope I'd gotten away with it. But after class as I was trying to ease along the side of the room, Sister called out, without even looking up from the papers she was marking, "Rita, would you please stay a moment? I'd like a word with you." In a panic, I looked around for Dolores and Marsha, but they were both making a blank-faced getaway. I froze, my books in hand.

Sister crossed to the door and latched it. No doubt about it, I'd be on the bus by sundown. I could feel my knees knocking together so hard I wondered if I was going to fall. Had she seen me sneaking off campus? Could she smell smoke on my fingers? I could taste adrenaline at the base of my tongue.

"You may sit down," Sister said cordially, as if she'd just invited me to stay for a chat, but I couldn't move, so after a moment she said, "Rita, can you tell me about what happened to your wrists?"

I neither moved nor spoke, and Sister came out from behind her desk and approached me, a mass of cloth of God. Still, her expression was benign. Behind her glasses, her blue eyes were open as a May sky, framed by long, thick black eyelashes. Sister Mary Angelice, too pretty to be a sister.

"Darling, what's going on? What's happening to you?"

The dressing-down I had expected I could have handled, or so I thought. But this kind concern unhinged me. It was if she'd torn a hole in an inner dam. First a stream and then a flood of grief came surging out of me. I began to sob, so hard it felt as if I was going to throw up.

"Rita, sit down," she said. She took my books and guided me to a desk, and then she put her arms around my shoulders and held me until my sobbing wrung itself out. She reached under her bib then and handed me a handkerchief.

When I could speak, I looked up at her and blurted it. "I just can't become a nun."

"A nun?" Her brow crinkled and she tilted her head. "Nobody ever thought you would."

I looked up at her, wondering if this was some trick. I'd seen those octagonal glasses before, but I couldn't remember where. And then I did. Mama used to wear them. Sister Mary Angelice smiled, revealing a goofy front tooth. "Where ever did you get the idea that we expected you to become a novitiate?" She sat down in the desk Marsha had vacated, gathering the folds of her skirt.

"Daisy said—" I couldn't finish. I felt too stupid. All I could do was stare at her habit, marveling at how skillfully the fabric was draped, so that it flowed without revealing any buttons or pins. How did they manage the thing when they went to the bathroom?

"Rita, are you in there?" she said, gently grasping my hand. "Honey?"

"Well, my aunt told me that you'd all agreed . . ." I stopped, realizing how dumb it sounded. Often, just like Daisy, I believed the dumbest things. Was I going to grow up and be exactly like her?

I couldn't look Sister Mary Angelice in the eye, so I stared past her to the front of the room. The black crucifix above the blackboard had a coating of chalk dust. "Well, you all were so nice and . . ."

She gave me a wry grin. "Is that supposed to mean we aren't generous with girls who don't have a religious vocation?"

"I didn't mean that."

"Rita, the motherhouse is in Chicago. Mount Saint Gertrude isn't a convent, only a school." I wanted to explain that I thought they made nuns here too, but then it dawned on me. Daisy had probably figured I'd toe the line extra hard if I believed I had to present myself as a novitiate. She never understood that forcing things always backfired.

Sister Mary Angelice got up and went to the front of the room, where she poured me a glass of water from a pitcher on her desk. I drank it down gratefully, in spite of the soapy taste. Then she said, "So, are you going to answer my question?"

I stole a glance up at her and saw that she was looking at my wrists. I turned them over.

"Rita, it's pointless to try to hide here. I noticed those the day you came. Do you think we are blind? That we can't recognize a lamb in need of shelter?"

I really didn't care for that particular choice of livestock. "It was an accident," I said. "I fell and hit my arms on a barbed wire fence. Daisy and me were fixing fence." I knew how lame it sounded, but how could I let anyone know the depth of my self-hatred? If she trotted out that prattle about how Jesus' blood had never failed me yet, I would truly explode. But she didn't.

"I see," she said, when I still refused to look at her. After a minute, I could feel her let it go. She stood up and patted me on the shoulder. "The information we are covering now is going to be on the college entrance exam next year. I believe you are college material, but if you wait to learn it, you'll be sorry. If you need my help, dear, you'll have to let me know." And then she walked to the blackboard and picked up the eraser and began clearing the board.

Her acceptance of my choice knocked me off balance. Instantly it made me want to confide in her, but I had no idea how to get out of the corner I had backed myself into. And by now she was seated at her desk again, immersed in student papers. I picked up my books and left, befuddled.

That afternoon, when it was time for Marsha and me to slip through the fence, I ran into Maple. Impulsively, I followed her to the library instead. Immediately the sense of doom that seemed to follow me around night and day lifted. I remembered how much I had loved the library back in Steamboat. I tackled pluperfect declensions of Latin verbs with an intensity that made my eyes cross, realizing that I was going to need every single minute remaining in the semester if I was going to pull up my grades. How ill-equipped I was for the outlaw life!

"Would you like to come home with me for Thanksgiving?" Maple asked later that week. I had just assumed I would spend the holiday at school. It had never occurred to me that the sisters might let me go elsewhere. "Can't see why not," said Sister Mary Angelice when I worked up the courage to try it out on her. "All you have to do is have her parents contact Sister Mary Regina."

As I waited with Maple for her parents, I was excited but apprehensive. Clearly Daisy's opinions about urban blacks hadn't prepared me to meet the family that had produced Maple Rollins. Would my country manners humiliate me? Would the prejudices Daisy had ingrained in me show? Nothing she had ever told me had prepared me for the handsome, calm man who walked in the door, with mushroom-colored trousers and a heathery taupe sweater. Maple's mother wore an ivory wool coat with a pale mink collar, and ivory shoes, gloves, hat, and handbag Daisy would have found unbearably vain. She wasn't exactly pretty, but she was entirely poised. "Maple has told us so much about your fascinating background," she said. "I can't wait to talk to you." That flummoxed me entirely. I couldn't think of a thing I'd want to tell this perfect family about my life with Daisy.

The car smelled new, and I was shocked to learn that they bought a new one every fall. The surprises didn't stop there. Their Denver home had been "designed," and by a black architect. It was fancier than houses in Steamboat even. There was a den, a living room, a dining room, a breakfast nook, a pretty kitchen in white and cobalt blue, and three bathrooms. They had a cleaning woman and a Japanese gardener who kept their hedges trimmed as tightly as a poodle's fur. These were roles my uncles and aunt performed for white people. The yard had been landscaped—another new concept, which I discovered meant you didn't plant onions alongside the tulips. In fact, they didn't plant onions at all. If they wanted an onion, they went to a store.

The den contained African masks, beadwork from people called Zulus. The only Africans I had ever seen were in Tarzan movies or the worn-out copies of *National Geographic* in the Steamboat school library, where the ladies with their naked boobies enabled the boys to look down upon the savages they were contemplating on the one hand, and connect to what was savage in themselves using the other.

Maple's ivory-colored baby grand piano dominated the living room. Not only did Maple play, better than I, but her father sat down and jazz spilled into the air, a jumble of notes with the melody buried inside. Maple's mom did something important in the mayor's office. Her uncles were lawyers and doctors. There were pictures on the mantel, and the pantry was crammed with food.

Maple's room glowed like a Russian snow palace, with lacy curtains that matched the frothy white bedspread. A herd of teddy bears frolicked among the frilly pillows, and she had her own princess phone. Bookshelves lined one whole wall, and there was a walk-in closet full of clothes. She had a little desk by a window that looked out on the pool, and an entire encyclopedia all to herself.

Maple's home demolished everything I had been taught about black people. Daisy saw black people as irrelevant in the greater scheme of things, and so she drove me to succeed in a white world. It had never occurred to me that there were black people who lived their lives without constantly using whites as a reference point. At Maple's I could detect no hint of the pernicious race consciousness that seemed to infect my family.

I couldn't think of a thing to say to them. Stupefied by the quality and quantity of available food, I ate my way through the weekend in a mute daze, and was relieved when Sunday night found me back in my little pen where I could unpack my thoughts. Did I want to live as they did, tidy and tight? I could see I was more free-range in spirit, but I could see the virtues of central heating.

The Thanksgiving experiment was so enlightening that I readily agreed when Marsha asked me home to Fort Collins for Christmas. Although I didn't know the exact word, I was starting to think like an anthropologist, and I wanted to see families at work.

Marsha lived in a small town on the eastern slope in that flat part of Colorado with its ass backed up against Nebraska. Their house sprawled up three stories, and her cheery, bleary parents barely noticed I was there because her four brothers' friends had piled in as well. Mother, father, kids were all freckled, their blue eyes shot through with veins as red as the stripes on an American flag. There were cases of Schlitz stacked on the back porch. Empty cans spilled out into the yard.

Christmas dinner turned out to be an adjunct to football. The bird got almost done. Sweet potatoes, cranberries, relish— everything came out of a can. Marsha slept a lot, and I felt like a goose with glasses that had flown in and landed at a tailgate party. They screamed themselves hoarse in football jargon— "offsides" and "Hail Mary pass" and "backfield in motion"— which I understood not at all. When I stepped outside on the porch to find some balance, the razor blade wind that raked the prairie seemed so bleak that I figured I might as well go back inside and see if somebody could teach me the point of the whole thing.

Marsha did have cute brothers. Greg, the eldest, had flared ears and a cherried-out Chevy Impala with a big engine and flared fins. "Let's go to the lake," he said later that night, jiggling the keys. I piled in the back with Marsha and another brother, and the other three crammed themselves in front. The sporadic patches of black ice on the highway daunted Greg not one bit. Out through the rolling hills at a hundred miles an hour he floored it, so that the car went airborne at the top of every rise, taking my sloshing, beery stomach with it. Finally, I had to speak up. "You gotta let me out. I'm sick."

I went as far away from the car as I could so they wouldn't hear me throwing up. When I climbed back in, we sped on as it nothing had happened. Our destination turned out to be a frozen pond, which brought back terrifying memories of my winter fishing expedition with Daisy. I had never heard of "donuts" but Greg was happy to enlighten me: You got up some speed on the ice, then slammed on the brakes, causing the car to spin out. Of course, my stomach spun out too.

On the way back to school, I confided in Marsha how cute I thought Greg was. "He's so dreamy," I said, as we opened our wax-papered bologna sandwiches. Marsha grew uncharacteristically quiet. "That's funny," she said at last. "I thought you'd prefer a colored fella." In the months I'd known her, over all the shared cigarettes and pitchers of beer, that line had never come up. But now here it was, when I least expected it.

Back at school, I realized that my relationship with Marsha hadn't been a friendship so much as an alliance based on binge drinking. She had done me a favor really, singeing my fur just before I got well and truly burned. Though we continued rooming together, it was as if our trains had switched tracks. I locked into my schoolwork seriously, and slowly, with Sister Mary Angelice's help, I managed to turn my algebra grades around. Marsha hardly noticed. She was running with Dolores Smith now, and gaining an alarming amount of weight. After Easter recess, her cot was empty. Explanations from the nuns that Marsha's family needed her home only confirmed the whispers that had been going around—she'd been sent home pregnant.

~~~

Daisy, enormously relieved that I had made it all the way through my junior year, wasn't taking any chances. Summer was to be spent in absolute isolation. No Perry Mansfield, not even once. With so much time on my hands, I managed to finish *The Rise and Fall of the Third Reich*. It chilled me to realize how easily people could submerge themselves in a mob, and how easily a mob could turn murderous.

One boring night I got it in my head that I could go on an expedition to visit my father. I still heard from him at Christmas—he'd send a shirt, or some shoes that didn't fit, always accompanied by a card his wife had signed. Daisy didn't resist, although she did sigh and say skeptically, "Well, if he'll send you a ticket . . ." To the surprise of both of us, he did.

I boarded the bus with adventure fever. California glimmered still in my imagination—"Hollywood be thy name." I had no idea that my father lived nowhere near Tinseltown. This time, instead of traveling toward Denver, the bus headed west toward Utah, a fearsome state full of the tribe of Mormon. I didn't even get off the bus in Vernal. I loved looking at the other passengers, though, and I kept changing seats so that I could peer in on their lives. I recognized the nut-colored migrant workers on their way from the peach orchards of Grand Junction to the California orange groves. I admired the patience of the tired mothers with fussy babies clinging and drooling on their laps and older boys rooting around in the ashtrays for butts, pulling on the tattered upholstery. In a year's time, I wondered, where would I go? What could I do? I seemed to fit in nowhere.

After three nights and two days, the bus groaned into Oakland. I spotted Daddy and his family at the gate instantly. He wore his straw cowboy hat at a silly, jaunty angle, and beneath that he wore a wig, I assumed to hide his kinky hair. But his face was handsome and his build was still that of a man who knew how to sit a horse.

Marge, his wife, had her hair up in pin curls and a thick layer of fuchsia lipstick on a little cupid-bow mouth. But Randy and Joyce were the ones I wanted to see. Randy, no more than ten, looked exotic, almost Italian, while Joyce's skin glowed pale and creamy. I was pretty sure she was a year or two younger than me, but she looked older. I saw that they were straining to see me through the smoky glass, and suddenly my fear caught up with me. They were the ones Daddy was providing for. I was the interloper.

The Oakland air landed heavy and wet in my lungs, making me feel like I wasn't getting enough to breathe.

"Uh, uh, Rita, how ya doin?" I had forgotten my father stammered. "It's nice to see you."

Marge, with the top-heavy build of a pigeon, looked as if she might fall forward. The two kids had remarkably beautiful eyes. We clearly only wanted to gawk at each other and compare features, but the adults bundled us out of the station in a whirl and piled us into the car. We drove past many squat houses too close to each other. Something about the town seemed tired and mildewed, as if the settlers of this region had used up their reserves getting this far and lost heart.

"You hungry? We can go past the Colonel on the way to the house," said Daddy, with a kind of forced joviality. Daisy would have been shocked at the idea of spending money on eating out, but I was delighted.

"How's Daisy getting along?" asked Marge, as if we were neighbors living down the road. I thought about the bitterness that had raged between my father and Daisy, and the air felt too close. "Fine," I said, realizing how little I would be able to say among them.

Back at the house, the chicken that we'd bought at the Colonel's tasted as if all the flavor had been boiled out before it was fried. The biscuits were half done and the potatoes made from dehydrated flakes. It was all very modern, but there was something missing. There was something absent from the house too. It wasn't just knowing that my father, the master carpenter who could build anything, still rented his home, as Daisy delighted in pointing out. It was that the house contained no books, not a single one.

The empty feeling persisted the next day, hard as they all tried to make me welcome. They were cheery and careful to do the right thing. They took me to San Francisco to see Chinatown and the wharf. I had never seen the sea before, and it struck me as remarkably smelly.

Randy was besotted with Elvis Presley. When I asked him if he wanted to play the guitar, I realized it had never occurred to him. He didn't want to be a musician. He wanted to be Elvis. He and his sister and my father and Marge watched television every single night of the week, even though they lived in a city with dance companies, theater groups, opera and other music of all kinds, but it never occurred to them even to read the part of the paper that talked about those things, let alone to get out and see for themselves. The best and the worst I could say about them was that they were nice.

It began to dawn on me, in brief little shafts of light, that maybe marrying this white woman was as much of an accomplishment as my father could muster—that although race was never mentioned, it was as present here as in Daisy's house.

"Your mama and me would have stayed together if it hadn't been for Daisy," Daddy said one afternoon when we drove alone to buy milk and bread. As I listened to his complaints about her meddling, I had to admit they sounded like the Daisy I knew. But I had to admit something else as well. Daddy would never have found me a German tutor or a piano teacher, or gotten me into private school. And though I would never have admitted as much to my aunt, I saw what she meant about my father. He wanted little more than to go to his job, come home, and have some peace. He had no interest in having a "public," the way Daisy did. He did not need to be known for his wisdom, his charity, his heroic contributions to family and community. He liked to work with wood, and that was that. And much as it pained me to acknowledge it, I knew I fell into Daisy's camp: *Do something, goddamn it. Be somebody.*

I had flirted with the idea of asking Daddy to let me stay with them. I certainly hadn't expected to feel sorry for my father, with his stammer and wig and silly hat. I had clung to the fantasy of him as a powerful cowboy who had hightailed it out of Dodge to where the pickings were better. But by the time the visit came to an end, I was ready to go. I realized that the hollow place I had reserved for family was going to remain unfilled.

In the meantime, I was Daisy's maniac, and I had things to do.

~~~

When I returned to Mount Saint Gertrude in the fall, I unpacked my suitcase in a bigger room. I was a senior now. I recognized the smell of the floor wax, the echo of girlish laughing from one end of the dorm to the other. I looked forward to the stimulation of classes and the company of people my own age. I had only nine months to go before I graduated, and I knew that with the help of Sister Mary Angelice, I could make it.

I caught up on gossip. Dolores Smith hadn't returned from Oklahoma. Over the summer, Mary Jo had started dating a football player at CU—one of those exotic, dangerous creatures who were completely off-limits to us. Mary Jo interested me anyway, because she wasn't even Catholic—she had come to Mount Saint Gertrude solely for the superior education—and this bit of news only added to her allure. I didn't even recognize Janet. She had gone on a grapefruit diet and lost thirty-seven pounds. I did embroider my visit to my father a bit, chatting up the glories of Telegraph Avenue in Berkeley.

After I had hung up my uniform and arranged a picture of Jiggs next to Jesus and Mary, I went looking for Sister Mary Angelice. Coming down the stairs, I could hear Sister Mary Regina's tinkling laughter. I had not lost my reflexive fear in the face of our tall, austere mother superior, but at least I could talk to her now.

After an exchange of the expected pleasantries, I asked her where Sister Mary Angelice was. Her smile never wavered. "She's been called back to the motherhouse."

I received the news physically, as if I had been kicked. "Does that mean she isn't going to be here at all?"

She looked at me blandly, her hands under her bib. "I rather doubt it. But we are all here to help you, dear. Now run along to the assembly."

"Is she ever coming back?" I persisted, against my better judgment.

"One doesn't question the wisdom of Mother Church, dear. We live a life of service." I hated it when they trotted out those platitudes, as if nobody noticed Mother Church was run by human beings.

~~~

The following weekend, the entire academy was to have a picnic atop Flagstaff Mountain. We set off around noon, but soon we were all spread out along the road that wound up the mountain. With my sturdy mountain legs, I got to the top before the others. In the parking lot where we were to light a grill, a white convertible full of college boys idled.

"Whoa, check out the chick," they howled, stomping the floor of the car. I flushed and scooted over behind a boulder.

"Is that a Gertie girl behind that rock?"

Exhilarated, terrified, I didn't say anything, but I sneaked a peek to take stock. Four guys sitting on the shoulder of the backseat, waving cans of Coors, the driver in sunglasses with a mangled straw hat. But the one riding shotgun made my eyes pop—he had skin the same as mine, and turquoise eyes. Even at thirty feet, I could feel them trying to pierce the stone.

"Come out, come out, wherever you are," yelled the driver, popping the clutch, making the car lurch over to the edge of the rock. He looked drunk enough to drive the car off the mountain as a joke.

The boys began piling out of the car. Panic rose in my throat. I looked behind me. Just past the railing was a sheer drop of hundreds of feet, and there was nowhere else to turn. And then the blue station wagon crested the rise, and Sister Mary Regina, Sister Mary Daniella, and Sister Mary Joseph pulled up next to the white convertible. I was relieved to see them—and disappointed.

"Oh shit, the penguins," the driver yelled, loud enough to be heard in Denver. "Let's get out of here."

But the black boy jumped out of the car and was already loping toward my hiding place. He came around the other side and there we stood, face-to-face. I couldn't say anything, but he did. "Hey, Gertie girl. You belong to me." I tried to bristle at his presumptuousness, but I was completely bowled over. "What's your name?" he asked. I went mute as a country mouse and bolted past him, terrified the nuns would find me before I could extricate myself. But as I passed he said, "We'll be around all day. I'll meet you halfway down." And then he was gone.

While the rest of the girls came up the hill, I helped to start a fire in the barbecue grate, my mountain skills finally of use, but my mind was elsewhere. "Rita's gone fishing again," said Sister Mary Daniella, and I looked up to find everyone at my picnic table staring at me and I blushed a deep Indian red.

When it was time to hike back down, the line sprawled out as it had before, and again I took the lead. He was waiting for me, three-quarters of the way down. His group, splayed around a picnic table, appeared drunker now, and they had a contingent of coeds in cutoffs sitting in their laps and perched on the car hood. But this boy seemed sober, and very focused on me.

"I told you I'd find you again." He towered above me, with eyes nearly the color of a robin's egg. He held out a piece of notepaper. "I'm David. I live in Denver. This is my number." He pressed it into my hand. "You have to call me. Tonight." And then he rejoined his friends at the picnic table, as if we had never met.

I continued on back to campus, with no plans to call him at all. But I was delighted at the prospect of having my own personal intrigue to report. This time I would have something to contribute when Mary Jo started carrying on about the varsity sweater Mr. CU had given her. I thought of David in religion class, during lectures about the mortification of the flesh. I toyed with my secret while the lay social studies teacher droned on about a country halfway around the world called Vietnam. The French had abandoned it or gotten thrown out or something. And the top part was Communist and wanted the bottom to be Communist and we had troops there now to stop the dominoes.

I wasn't a complete fool. I knew there was something far too calculated about his come-on. But a handsome boy had found me pretty! When I pictured his face in my mind, I skipped over the skimpy mustache and the sprinkling of acne. And I ignored the high-handed way he'd told me to call him, and his disregard for the risk I'd run by doing so. How exciting to fantasize about pursuing an adventure that was my own—something not dictated by men and set down in stone in Italy in the year 276 A.D. Something immediate, my own, fresh. I could only imagine Daisy's reaction if she learned I'd gone with a black boy. Ha!

I kept the note in my sock drawer for three weeks. And then one night it occurred to me that I would never know what he was like if I didn't make the call. I went down to the phone booth and put a coin in the slot.

"Yeah," he answered, his voice low and cool. I was attracted to it even as I registered how phony it sounded. Through the window, I could look into the mother superior's office, read the plaque next to her desk light that read SISTER MARY REGINA, BVM.

"When am I going to see you?" David asked. His aggressive directness overwhelmed me. He didn't ask if he could see me. He demanded to know when. In my confusion, I couldn't sort my thoughts. I had had little experience standing up for myself, and I didn't want to be impolite.

"Well, I can't get away so easily," I told him. "I'm not even supposed to be talking to you."

"You're breaking my heart, sweetheart."

That sounded like something out of a movie. "Oh come on," I said, but I was flattered.

"Serious, I'm serious," he said. I studied the wood paneling of the booth, delighted. "Baby, I think about you all the time. Look, all I want is to have a cup of coffee with you."

"Well, maybe I could come visit my aunt and uncle and we could have coffee before I go to their house." Where had that come from?

He was on it. "There's this great little place just up the hill from the Brown Palace. You'd love it." And then he actually started singing. "Nothing you can do can make me be untrue to my girl." Even I knew how cornball that was, but he didn't stop there. "You have to stop torturing me. I love you."

Perspiration broke out on my brow. I couldn't believe he'd actually said that. Not to me. "I have to go," I said. I hung up and leaned my head against the wall of the phone booth, letting his words roll over me again. Someone had said it. *I love you.* Never mind that it was just words.

When I opened my eyes, Sister Mary Regina was looking dead at me. My eyes went wide. I opened the door. "How's your dear Aunt Daisy?" she said.

"Fine," I said, and dashed upstairs.

~~~

Finally, at the end of October, I set it up. I arranged to visit Uncle Billy and Aunt Helen in Denver for the weekend, and told them I'd be at their house by eight on Friday night. I got in around six and, following the directions David had given me, walked up the hill. I could smell the coffeehouse from half a block away.

Through the window I saw a group of people, all college age at least, dressed entirely in black—the uniform of cool people. Two girls wore short leather skirts with black tights and boots. The men wore turtlenecks, and draped themselves over their coffee in oddly static poses of boredom and gravity. I thanked God for giving me the foresight to bring jeans and a black sweater to change into at the bus station.

I opened the door and stepped inside. Bob Dylan was playing "The Times They Are a-Changin'." A little frisson of delight as I spotted David right away in a corner, hunched over a tiny cup of coffee like he was worried somebody might try to snatch it, a black beret sideways on his head. He didn't notice me right away because he was reading, and it crossed my mind that I could turn around and walk right back out. I was standing there, trying to decide whether to take off my glasses, when he looked up and flicked me the smallest smile, followed by a very restrained nod. I took off my glasses.

He jumped up as I walked toward him. "So, you made it." I couldn't gauge his expression, that abrupt manner. His nose shone a bit too much. I sat down across from him. I didn't know this man at all.

He leaned in to me, sly and slick, and whispered, "I have never seen eyes prettier than yours in my entire life."

I couldn't stand to look at him directly, for fear of shorting out all my circuits. "What are you reading?" I asked, desperate for something neutral to talk about until I could get my bearings.

"About the draft. I'm eligible. Don't want to go and fight no Vietcongoleses." He gestured toward the coffee bar. "So, what you want. Espresso?"

And then he jammed his leg against mine, hard. I didn't want to admit it scared me. I moved my leg away. "Excuse me," I said spying a cigarette machine. I knew how to do that. When I returned I occupied myself with the business of the cigarette pack, pulling the strip of cellophane, unfolding the foil, tapping the pack, pushing up the stair of smokes. I didn't know what an espresso was. "Can I have just a regular coffee?" What I really wanted was a glass of milk and one of the huge cookies I saw in a canister on the counter, but I didn't want to seem like a baby.

"Sure," he said. "It costs eighty cents." That shocked me. It had never occurred to me that he wouldn't buy my coffee. I could hear Daisy whispering in my head, "I told you so. Layin up in bed, livin off womens, and the womens let 'em do it." But I got up and paid for the coffee, because I didn't know what else to do.

"So are you a militant or a CU jock?" I asked, buying time.

"I am a black man, trying to get along." The self-pity sounded canned. But what did I know about black men?

In the background, a trumpet seemed to scream. "Who is playing that horn?" I said. "I've never heard anything like it."

"That Miles," he said. "Miles Davis. Come on. I want to stop by the house, let you hear his newest album."

"Only for a second," I said. I knew this was a really bad idea, but I couldn't think how to get out of it without being rude. And then we were outside, almost running down the sidewalk, which was covered in silver ice like steel plates. David put his arm around me, which didn't feel warm so much as aggressive. But I didn't want to let on, like some country bumpkin, so I said nothing.

The yard at his house had bare spots, as if it had been charred. He took me around the back, where there was no illumination at all, and opened a door. I realized he was taking me to the basement. How, I asked myself, had I managed to find a boy who would bring me to a place that felt and smelled just like home—the same moist cold, the same dense cement walls that entombed the air, deadened sound. "Be careful," he said as we walked down narrow wooden boards. With a profound sense of dread, I watched myself obey like a good girl.

But once he'd brought me down the dark stairway, he did not head for a phonograph or look for anything. That floaty, frozen-trout feeling started to come over me, like when Daisy was going to whip me. He turned on a tiny light that sat on an overturned crate next to a cot. He sat down on the cot and patted it. "Why don't you rest your coat?"

I did not want even to open my coat. But I was more afraid of looking afraid than I was of dying. When he patted the cot again, I sat down gingerly on the ratty pink blanket with its frayed ribbon. There was no record player, nothing at all that I'd imagined around him those nights when I had sat in the phone booth, coiling and uncoiling the telephone cord. The ceiling barely cleared six feet. In the far corner sat a pink washer and dryer. Clothes were piled on the floor, hanging from the overhead pipes. On the floor by the crate, a *Playboy* magazine featured a blond in a skimpy Santa suit that revealed her legs, stomach, and the tops of her pushed-up teats. I was confused and fascinated by a woman presenting herself like that. What about her reputation? Her salvation? What about her parents? Her neighbors? What if she knew that her likeness had wound up down here in this basement . . . like me?

"You like that picture?" David asked, grinning. I flushed with embarrassment so extreme that I just closed my eyes and hunched over. "Because if you do like it . . ."

He leaned toward me, way too close. Under the coffee, his breath smelled like onions, and his teeth had collected something yellow and unpleasant at the gums.

He put his hand on my knee. I jerked it away.

"So, where's the Miles Davis record?" I asked. "Because I can only stay five minutes." I scooted as far away from him as I could.

"Let me take your coat," he said. He kept advancing down the cot until I was pressed up against the iron frame, so close to the wall I could smell the cement. He pried off my coat with rough urgency. "I can do it," I said, a bit rudely. I still thought he was handsome, maybe even sexy, but I no longer liked him.

"Let me see your eyes," he said, easing my glasses off. "You got some Irish in you?"

"I kind of wish you wouldn't do that," I said, miserable now with panic. I reached for my glasses, but he put them high above his head. If I extended my arms for them, I would have to draw close.

"Where'd you get them pretty green eyes?" he asked. He held my glasses out toward me then, but when I reached for them, he switched them to the other hand, so fast I could hardly follow. I didn't like this game. I wanted to go home, but I didn't want him to get mad at me or think I was some kind of tease. Plus, I had vowed not to assume like Daisy that "men just like a dog, don't want but one thing." But what if he did want just that one thing? The memory came to me of two dogs hooked to-gether down by the well house and Daisy spraying them with a hose, spitting with fury at how disgusting they were. What had I gotten myself into?

Footsteps advanced across the ceiling toward an interior set of stairs I hadn't noticed earlier. David reached past me and turned off the little light. "Keep quiet," he said. I could only imagine what his parents would assume if they found me down here. Would they call the nuns? The cops? I jammed myself against the wall, holding my breath as a key turned in the lock and the doorway opened, sending a bleaching shaft of light down toward the end of the room.

"David," a commanding black woman's voice called down, "Who you talkin to?"

He jerked up. "Nobody, Ma." His voice croaked like he had eaten a frog.

"David. I heard voices. It's late. You spose to be 'sleep."

What time was it? I was supposed to be at my aunt and uncle's by eight. I tried to sit up.

He put his hand over my mouth and squeezed. His legs wrapped around mine so hard I thought my knee would break.

"I musta been talking in my sleep, Mama. Ain't nobody down here."

In the distance, a church bell tolled the hour. I tried to count the chimes, tried to inhabit the silence between them, because in those instants I was still alive. There was a safe place out there that did exist if I could just stay alive long enough to get there.

The woman sighed. "Well, all right then, baby. But you best cut out eating spicy food before you lay down to sleep."

"Good night, Ma." The door closed, taking the light with it, and the key turned. The feet walked back across the ceiling. David relaxed his hold on my mouth.

"Bitch." The softest whisper, but it felt like a slap.

I tried to jerk myself upright so that I could find my glasses. "I have to go."

"Oh, no, sweetheart. I'm going to give you exactly what you came here for." He pushed me down on the cot, pinning my hands above my head and laying his torso upon me. "Don't fight it." He stuffed my protest down my throat with his thick tongue. "You know you want it." He reached under my skirt and began to claw at my underpants, prying my legs apart. All those years of ballet and horseback riding were not enough to keep them clenched shut. With surprising swiftness, he thrust himself inside me. To keep from crying out, I studied the ceiling in the dim light coming from the top of the stairs. Why hadn't they wrapped the pipes? Didn't they know they would freeze in winter? Wouldn't they burst, water exploding the steel so that the life of the house spilled down and flooded the core of the foundation?

When he was finished, he threw back his head. For a moment he remained above me, motionless, panting hard. Then he opened his eyes and looked down at me, seeming surprised that I was still there. I felt like I was covered with soup. I tried to peel my cheek off the concrete wall where it had come to rest so I wouldn't have to look at him. An odd embarrassment pervaded the room then, as though something intimate had taken place.

With a great sigh, he rolled off me. The cold shocked me as I propelled myself up and away from his touch. I fumbled around, searching for my glasses, but I couldn't stand the close room any longer. I needed air. I grabbed my jeans coat and boiled up the stairs without my glasses or my underwear.

Outside I was received by a night sky so shockingly beautiful it made me cry. The smog of Denver had smeared the cloud ceiling the color of peach jam, and under it I made my way to a streetlight and finally to the corner, where there was a bus stop and a bench. I sat down and rocked myself, trying to think what to do. I couldn't imagine showing up at Billy and Helen's house smelling like a whore. But what if he came after me? I had to pull myself together.

A taxi pulled up to the stoplight. The driver, a little old bald man chewing a matchstick, looked over at me. I watched him take in my disheveled hair, my swollen, teary face, my missing sock. He reached across the backseat and opened the door. "You need a ride?"

I burst into tears then, and got in the cab. I started to tell him I didn't have any money because I had left my purse, but he waved his hand behind him to hush me and said, "Where to?"

I gave him my uncle's address and watched the tidy brick houses of Denver flow by and I wondered how I had managed to play out Daisy's scenario so impeccably.

~~~

There was only one piece missing from that scenario, but it fell into place soon enough. I returned to school and began to count the days.

My menstrual cycle had always run like a German railway, every twenty-eight days on the dot. But the day it should have started came and went. A constant rushing in my ears now made it impossible to concentrate on my schoolwork. Three more days passed, and nothing. A week and a half, two weeks, a month. I couldn't kid myself. Like a little snowshoe rabbit, I had hopped right up to the snare and placed my paw inside the loop. I seemed to be wired to self-destruct in the most spectacular and devastating way imaginable.

A month later, it started. Not my period, the puking. At exactly a quarter after nine in the morning, when religion class was well and truly settled in, a wave of nausea such as I had never experienced surged up in me. There in the bright buttery morning, I tore out of the room like the barn was on fire, leaving behind all thirty of my classmates, in their brown plaid skirts and blinding white blouses. My stomach empty, I washed my face, squared my shoulders in the bathroom mirror, and returned to class. No one seemed to notice.

The next day it happened again. And the next and the next, always at precisely a quarter past nine. Same railway, different destination. That part of me that always watched, way outside myself, said, "Well, you got it done. You'll be out of here by Christmas." And then one day a few weeks later, when my waistband was starting to feel snug, I started thinking about grapefruit. I thought about grapefruit all day and all night, and the next day, at great peril, I snuck off campus to the little store in the village. I had to have cold, juicy grapefruit. But not just grapefruit—peeled grapefruit. And not just peeled grapefruit, pink grapefruit with every single shred of skin removed, not one iota of white pith anywhere. And next to, but not touching, the glistening sections, I had to have sardines. And not sardines in oil or tomato sauce, but sardines in mustard sauce. And if the little store had none, I had to risk going farther into town, because I could not go an entire day without a can of sardines in mustard sauce and a peeled, sectioned, pink grapefruit. As my waistband tightened, my panic grew.

Although I had resolved I would never do it, I called him. I didn't know what else to do. And there was no one to talk to.

"Talk to me," he said. The sound of his voice made the hair on my arms rise.

"David, I'm pregnant."

I thought the line had gone dead. Finally, he spoke. "How do I know it's mine? I mean, you know, Gertie girls love a shot of hip."

~~~

Astonishingly, the home accepted me without challenging my moral worth. Their only request was that I hold in confidence the names of the other girls I met there. Sister Mary Angelice sent me a package every month. With so much time on my hands, I began to read again. *The Diary of Anne Frank* filled me with longing to get out and accomplish something. I learned to sew and made myself a little suit. Toward the end, with no money left and, I thought, absolutely nothing to lose, I decided to call my father. Sure, he'd send me some money. And by the way, what was I going to do with the baby?

"I am going to put it up for adoption," I said.

"How can you do that?" he asked, completely without irony. "How can you give away your own child?"

In June, the baby came.

They let me sit with him just once. He wrapped his tiny pink hand around my thumb. I told him, "I'm so sorry I have nothing to give you. Wherever they take you, you will be better off than you would be with me." I kissed his little head. His eyes were the blue of kittens' eyes before they get their true color. And then they took him away.

When I called Daisy from the depot in Steamboat, it was clear she was surprised to hear from me. But I was too shattered to arrange for anything except what was familiar. I stood outside the station and waited for her to come pick me up. Steamboat had once more donned its pretty summer dress, flowers and birds and blue sky. A sign across the top of Lincoln Street announced the rodeo next week. Wide-body pickups pulling horse trailers rumbled across the bridge toward the rodeo grounds. I was neither glad nor sorry to be home, I just needed to be where I knew what to expect.

The Jeep rolled up, but Daisy did not get out. When I got in, she headed back toward Strawberry Park, confounded into silence. She didn't ask what had become of the baby, but the question lay between us, unanswered. That was fine with me. I had accepted that the really profound matters between us could never be addressed.

"They still had lots of snow on Berthoud when we came over the Divide." I had finally learned the weather game.

"Not enough to stop folks getting over the Forty, is it?"

"No. The tourists can make it through."

My heart sank when she told me she had planned a dinner party for the nuns and the priest that night. I couldn't stand to watch her cook and serve, hovering behind them while they buttered their biscuits but unable to let herself sit. Mercifully, she had another plan for me.

"You can't be in the house with me now that you're eighteen. You can stay out back, but you got to get a job and pay rent, electricity. I can give you something to eat, you don't have to worry about that."

"Fine," I said, grateful beyond belief.

Out back, the old poultry sheds had sagged since they had first been tacked together on no foundation, but at least they were home. At the far end, Daisy was raising a flock of turkeys, and next to that the baby geese who were still too little to go to the pen. To the left of them the chickens roosted, and I got the final two rooms, one of them just big enough for a bed. I salvaged some lumber and bricks from a stack out back, and by nightfall I had my own little nest. I went down in the basement and brought out Gama's old coal oil lamp and found her quilt. It was the first actual solitude and safety I had had in years. It comforted me to hear the chickens murmuring through the night in their little poultry dreams.

Next morning I went to where the geese were roosting and plucked out a sleepy gosling. I tucked it under my shirt and felt it peep in contentment at the warmth of my stomach and the pulse of my heart.

The following week, Daisy had to make a run to the city dump, and it occurred to me I should go with her and see if I could find a chair. I found not only a chair, but a blue velvet couch and cast-off fruit crates for bookshelves as well. I bought some Rit color and dyed some gunnysacks for matching blue curtains. And every night after Daisy went to sleep, I would go and sit with Jiggs and listen to the creek and watch the travels of the moon and wonder where my baby was.

I was ashamed to be seen in town. Welfare to cover the cost of the unwed mothers' home had had to be approved in Steamboat, and I was sure the entire town knew what had happened to me. But I managed to find a job waiting tables at the Harbor Hotel on the graveyard shift, which I thought would be perfect, because the load would be light and none of my old friends would ever see me carrying my stain of shame. It seemed I had found something even I could succeed at.

I was correct about the job allowing me to keep a low profile, but not about how easy it would be to succeed at it. I'd had no idea how hard waiting tables could be, and I hadn't reckoned on the paralyzing effect of all I'd been through. I could barely speak to anyone or look anybody in the eye. My shame was so total people thought I was either crazy or stupid. When another waitress came barreling through a door with six hot plates, calling out, "Behind you," I turned around and crashed right into her, sending the entire order flying. And so it went, mishap after mistake, collision to crisis, in an unending nightmare that would have been slapstick funny if it were happening to somebody else. I would forget to fill the salt and pepper shakers for the Sunday morning rush. And if I remembered to refill the catsup and mustard bottles, I could not remember to clean them too. Half the time, I was so panicked writing down the orders that I couldn't read my writing after, and just put down what I thought seemed right. I who had mastered trigonometry could no longer perform simple arithmetic. When a line of customers waiting to pay began to look the least bit impatient, I stumbled over eight plus six. The more irritated they became, the more hopeless I got, until sometimes I'd just beg them to add it themselves. I just wanted them to get the hell out of the restaurant so that I wouldn't have to face them gawking at me in bewildered scorn.

My fantasy of calm nights when the load would be light turned out to be a joke too. Steamboat in the summer was a partying town now. At midnight the 3.2 beer bars shut down, and all the drinkers between the ages of eighteen and twenty-one came piling in to put straws up their nostrils and pour all the salt on the table and see if they could balance the empty shakers on their noses. I would look up and see twenty kids my age headed my direction and just want to walk back in the cooler and hang myself on a meat hook. At two in the morning, when I was just starting to get the tables cleared, the Pioneer closed down, and a hungry crew of honky-tonkers descended, demanding steak. Steak, that is, not gristle. Was I crazy? This was gray and tough as old catgut—and cold to boot. And how could it be that I got eight orders in, but forgot to put in the ninth? Only way I could make it better would be to set that cute little fanny on this good ol' boy's knee. It all seemed to come to a head when one "young feller" allowed how I looked like I'd "done been rode hard and put up wet." I dumped a cup of coffee in his lap and walked out.

All the next day I slept, and when it came time to go to work, I didn't even bother to ask Daisy to take me because I knew I'd lost the job. But around 10:30, the cook called.

"Where you at?"

"I thought I was fired," I said.

"Why?" he said. "Oh, you mean ol' Bob? That ain't no big deal. I'm sending the busboy to pick you up right now. We got the rodeo crowd." And so the summer stumbled on.

But each day, after a short sleep in the furious heat of midday, I would open an old book of Russian literature I still had from Whiteman. When I read the selection from *Anna Karenina*, something haunting and aching seeped into me. I particularly loved the dissolution in *The Cherry Orchard*, the same obsession with lost glory that held Daisy captive. The fact that I had mind enough to immerse myself in a play surprised me. I really believed my mind was Swiss cheese, incapable of any kind of through line at all.

And then I accumulated a suitor. Red worked the oil derricks, had lost his front teeth when somebody whacked him in the face with a beer bottle. He looked like a boiled crustacean wearing a haystack. He refused to believe that I had long since given up on life, declared that he was "sweet on me" and brought me a heart-shaped locket in a blue velvet box. He thought somebody who had just barely turned eighteen had her whole life ahead of her and by jiminy, he was determined to set me straight. He announced that every single night he was going to come into the Harbor and drink up all the Tabasco sauce until I'd give him a real smile. And so there was no hot sauce for the morning shift, day after day after day well into the middle of August. I looked at Red, long and hard. And, in spite of his rough, grease-stained hands, there was a kindness in him that I knew would wear me down. I could see myself moving into his trailer and having seven kids in no time. I'd learn how to down boilermakers and play piano at the Pioneer and tell every drinker who would listen what a dancer I might have been.

Somehow, in spite of everything, I accumulated a sockful of change. After a month, I had almost eighty dollars. I bragged about this at supper one evening to Ernest and Daisy. Daisy was extremely impressed. "Well, it's a surprise, I'll tell you that, to see you working with the public. Yes sir." I blushed with pleasure.

Next day Ernest caught me on the way down to the creek. "Say, Rita, would you mind if I borrowed your tip money? I'll pay you back in a week."

"Ernest, I need this money for college. You got to promise to give it right back," I said. It was a ludicrous idea—me going to college when I had left four high schools without graduating from any of them. I had not applied anywhere and, in fact, hadn't really thought about college until that moment.

"Oh yes. You'll have it next week, you'll see."

Knowing I was beyond saving, I went up to my little shed and dutifully retrieved my sock of change from its hiding place in the eaves. I took it down and put it in Ernest's hand. That night he was on a bus to Denver to play the dogs.

The next day, after work, I didn't go to bed. Instead, I went into Daisy's house and called every college admissions office in Colorado, begging for somebody to take me. I knew it wouldn't work, couldn't happen, but what did I have to lose?

Ernest did not come back the next week, nor did he return my money. He was on a winning streak, so he followed the dogs to Arizona, because he'd always had really good luck there.

None of the colleges called me back, but I was put on official warning that if I could not learn to manage my hot sauce any better, the restaurant would have no choice but to look for someone who could.

The first of September came. On the seventh, I got a call from Mesa Junior College in Grand Junction. They had space. Did I still want to come? I screamed into the phone.

I took the Trailways there, of course, and when I arrived in Grand Junction, Daisy had left a message for me to call her.

"Western State said come down to Gunnison."

I didn't even unpack. Just went on down to the station.

I didn't know it yet, but there was a music school there, with practice rooms where I could go and play the piano for hours at a time, and I was going to make my way through Chopin's polonaises and the Warsaw Concerto and *Slaughter on Tenth Avenue*. I was going to learn to ski for real, and to fence as well. I was going to discover a library the size of a city block. Some days I would go there and close my eyes and settle into the blond wood corner and devour the books like a woman starving. I would close my eyes and pick a book, any book. One day it would be Viktor Frankl's *Man's Search for Meaning*. Another day it would be a book about the life of bees. I'd find a word portrait of Daisy—*narcissistic personality*. And a picture of my own psyche in that experiment with the little monkey clinging to the mannequin made of terry cloth and wire.

But I didn't know any of this when I got on that bus the final time. I only knew that when I heard that same old grinding of gears, this time it was taking me some place I wanted to go.

ACKNOWLEDGMENTS

Some names and identifying details in my account have been changed to protect the privacy of those I write about, but otherwise I have been true to people, places, and events as I remember them.

Of course, the road from memory to book is a long one, and I owe much thanks for help along the way. On an absolute fluke, I applied to the Cottages at Hedgebrook in 1991, where I was given a writing cottage for six weeks. Although this was an auspicious beginning, my stamina still flagged, and this manuscript languished in grocery bags in Jackie Gorman's garage for a couple of years. Then Alison Gold got it to *O, The Oprah Magazine,* and Oprah Winfrey bought a version of what would become chapter 5. With blinding speed, Ellen Geiger, my agent and friend, got the book sold. My writing friends Jonathan, Russell, Barry, Birute, Jackie, and Vickie listened to ten thousand interminable drafts. The incomparable Janet Fitch, David Francis, Julianne Ortale, and Sheron Steele critiqued and supported the work with invaluable compassion and clarity. Anne Stockwell was the kind of partner one dreams of—brilliant, funny, an incredible editor and friend.

But this book would never have happened without Rebecca Saletan. Traipsing along behind her from publisher to publisher, I made the right choice. And I knew it from day one.

Those of you who have saved my life repeatedly, thank you for your kindness, vision, and generosity: Abigail Bok, Alyssa Selby,

Acknowledgments

Amy Bloom, Amy Gross, Anna Abreau, Annie Juchat Bear Meyer, Ben Metcalf, Birute Serota, Brother David Stendl-Rast, the Buddy Werner Memorial Library, Carol Stanley, Max Haas, Cass Lyons, Charley Bates, Charlotte Perry, Cheryl Cabasso, Chris and Tracy Mallett, Cindy Radtke, Cynthia Grayson Pearson, David Francis, Darold Wax, Devin Robinson, Dolores Valdez, Gregg and Stephanie Ross, Henry Finder, Henry Louis Gates, Jean Hardesty-Radecki, John Q. Cope, John Fox, John Whittum, Jonathan Authur, Joy Johannessen, Julianne Cohen, Judy Wieder, Kate Braverman, Kathleen Wakefield, Kay Uemura Henderson, Kit Rachlis, Leonard and Susan Post, Lorene Workman, the Los Angeles Public Library, Lynne Overmeyer, M. G. Lord, Marie Aguirre, Marie Sandelin, Marilee Saxe, Mary Agnes Catherine Muldoon Morris, Maybell Chotvac, Michael Vasquez, Michael Ventura, Nancy Nordhoff, the Cottages of Hedgebrook, Pat Towers, Portia Brown, Portia Mansfield, Renata Karlin, Richard Gartland, Sam and Anne Hardy, Sherrill Soliz, Sister Mary Angelice, Stacia Decker, Steven Carter, Suzanne Elussor, Tannis Kobrin, Robert Reid, Victoria Pynchon.

I hope that the many others whose kindnesses to me I have neglected here will forgive me the oversight.